Applied Statistics in Social Sciences

Emilio Gómez-Déniz
Department of Quantitative Methods
University of Las Palmas de Gran Canaria, Spain

Enrique Calderín-Ojeda
Department of Economics
University of Melbourne, Victoria, Australia

CRC Press is an imprint of the
Taylor & Francis Group, an **informa** business

A SCIENCE PUBLISHERS BOOK

First edition published 2022
by CRC Press
6000 Broken Sound Parkway NW, Suite 300, Boca Raton, FL 33487-2742

and by CRC Press
4 Park Square, Milton Park, Abingdon, Oxon, OX14 4RN

© 2022 Taylor & Francis Group, LLC

CRC Press is an imprint of Taylor & Francis Group, LLC

Reasonable efforts have been made to publish reliable data and information, but the author and publisher cannot assume responsibility for the validity of all materials or the consequences of their use. The authors and publishers have attempted to trace the copyright holders of all material reproduced in this publication and apologize to copyright holders if permission to publish in this form has not been obtained. If any copyright material has not been acknowledged please write and let us know so we may rectify in any future reprint.

Except as permitted under U.S. Copyright Law, no part of this book may be reprinted, reproduced, transmitted, or utilized in any form by any electronic, mechanical, or other means, now known or hereafter invented, including photocopying, microfilming, and recording, or in any information storage or retrieval system, without written permission from the publishers.

For permission to photocopy or use material electronically from this work, access www.copyright.com or contact the Copyright Clearance Center, Inc. (CCC), 222 Rosewood Drive, Danvers, MA 01923, 978-750-8400. For works that are not available on CCC please contact mpkbookspermissions@tandf.co.uk

Trademark notice: Product or corporate names may be trademarks or registered trademarks and are used only for identification and explanation without intent to infringe.

Library of Congress Cataloging-in-Publication Data (applied for)

ISBN: 978-0-367-64204-4 (hbk)
ISBN: 978-0-367-64205-1 (pbk)
ISBN: 978-1-003-12343-9 (ebk)

DOI: 10.1201/9781003123439

Typeset in Times New Roman
by Radiant Productions

Preface

This book reviews some of the more relevant statistical distributions in the literature and their application in several fields of social sciences, including actuarial statistics, finance, income distributions, regional geography, tourism, etc. Undoubtedly, the tool of probability distributions, discrete and continuous, univariate and multivariate, constitutes the fundamental element of work in all these settings. Hence, the first chapter of this book aims to know and present the most important statistical distributions used in those scenarios. Readers with extensive knowledge in the field can ignore this chapter if they are interested in reading later chapters. Therefore, we would like to emphasize that this text is designed in a self-contained way so that those readers who wish to refer to a specific area for those readers who are interested in a particular area can refer directly to it without reading previous chapters.

In the second chapter, we study the application of the statistical distributions described in the first chapter in insurance and finance. Here, different methodologies to deal with aggregate claims in an insurance portfolio are discussed. Furthermore, we provide several mathematical methods to calculate premiums and risk measures in insurance, reinsurance, and finance. Finally, risk ordering is considered in the final section of this chapter. In 2019, the tourism industry constituted 10% of the world's gross domestic product. For that reason, in the third chapter, we consider using the more relevant probabilistic families in tourism. We will focus on variables such as the length of tourist stay at holiday destinations and the expenditure per tourist. This book's final chapter briefly addresses four areas of economics that have attracted much interest in recent decades from the research community. These include stochastic frontier models, models in Geography in an urban agglomeration analysis, duration models, and income distribution models.

A list of exercises proposed at the end of each chapter is included. We encourage the readers to complete this set of problems and prove and obtain the results given in each question. These exercises are, on many occasions, the results of the derivations implemented in each of these chapters. Therefore, they constitute in themselves an invaluable source of expansion of the knowledge acquired in the corresponding chapter.

Whether or not engaged in research, we hope that readers find this book a reliable source of information. Without a doubt, we will also appreciate your sending us all the errors that it may contain.

Contents

Preface   iii

List of Figures   vii

List of Tables   ix

1. Basic Statistical Distributions   1

    1.1 Introduction   1
    1.2 Univariate discrete distributions   2
        1.2.1 Bernoulli distribution   2
        1.2.2 Binomial distribution   3
        1.2.3 Moment and probability generating functions   4
        1.2.4 Poisson distribution   6
        1.2.5 Negative binomial distribution   11
        1.2.6 The geometric distribution   13
        1.2.7 Logarithmic distribution   15
    1.3 Univariate continuous distributions   16
        1.3.1 Normal distribution   16
        1.3.2 Lognormal distribution   17
        1.3.3 Gamma distribution   18
        1.3.4 Exponential distribution   20
        1.3.5 Weibull distribution   21
        1.3.6 Inverse Gaussian distribution   23
        1.3.7 Family of Pareto distributions   25
        1.3.8 Classical Pareto distribution   25
        1.3.9 Pareto type II or Lomax distribution   27
        1.3.10 Beta distribution   28
    1.4 Deriving new distributions   29
        1.4.1 Mixture of distribution   29
        1.4.2 Composite models   33
        1.4.3 General composite models   35
    1.5 Multivariate distributions   35
        1.5.1 Bivariate Poisson distribution   35
        1.5.2 Bivariate Poisson distribution. An alternative parametrization   36

Contents v

	1.6	Multivariate continuous distributions	38
	1.6.1	The multivariate normal distribution	38
	1.6.2	Bivariate exponential distribution	41
	1.7	Criteria for model validation	42
	1.7.1	Hypothesis testing	42
	1.7.2	Other measures of model selection	44
	1.7.3	Graphical methods of model selection	44
Exercises			46

2. Statistical Distributions in Insurance and Finance 49

	2.1	Introduction	49
	2.2	Individual and collective risk models	50
	2.2.1	Individual risk model	50
	2.2.2	Collective risk model	52
	2.2.3	Compound Poisson distribution	55
	2.2.4	Compound negative binomial distribution	55
	2.3	Classes of discrete probability distributions	56
	2.3.1	The $(a, b, 0)$ class of distributions	56
	2.3.2	The $(a, b, 1)$ class of distributions	56
	2.4	A recursive expression for the aggregate claims distribution	59
	2.5	Premium calculation principles	62
	2.5.1	Examples	62
	2.5.2	Properties of premium calculation principles	65
	2.6	Risk measures	67
	2.6.1	Value at Risk (VaR)	67
	2.6.2	Tail Value at Risk (TVaR)	69
	2.6.3	Conditional Tail Expectation (CTE) and Expected Shortfall (ES)	70
	2.6.4	Properties of risk measures	72
	2.7	Reinsurance	73
	2.7.1	Type of reinsurance	74
	2.8	Comparing risks	77
	2.8.1	Stochastic dominance	77
	2.8.2	Stochastic dominance and stop-loss premiums	79
	2.8.3	Stop-Loss order and Stop-Loss Reinsurance	82
Exercises			82

3. Statistical Distributions in Tourism 85

	3.1	Introduction	85
	3.2	Data	86
	3.3	The length of stay variable	90
	3.3.1	Models	92
	3.3.2	Numerical illustration	97
	3.4	The expenditure variable	98

	3.5	Compound models	102
		3.5.1 The compound Poisson model	104
		3.5.2 The compound positive negative binomial model	105
	3.6	Bivariate model	108
		3.6.1 Some methods of estimation	111
	3.7	Generalized additive model	116
Exercises			117

4. Statistical Distributions in Other Fields — 121

	4.1	Introduction	121
	4.2	Stochastic frontier analysis	122
		4.2.1 The general model	124
		4.2.2 The normal-exponential model	124
		4.2.3 The normal-half normal model	126
		4.2.4 The normal-truncated normal model	127
		4.2.5 An example	127
	4.3	Geography: The size distribution of cities	130
		4.3.1 The composite lognormal-Pareto	132
		4.3.2 Data	134
		4.3.3 Numerical results	135
	4.4	ACD models	138
		4.4.1 The general model	140
		4.4.2 Specific models	140
		4.4.3 Extensions	143
		4.4.4 An empirical example	145
	4.5	Income	147
		4.5.1 Basic elements	149
		4.5.2 Inequality measures and population functions	150
		4.5.3 Lorenz ordering	151
		4.5.4 Estimation	152
		4.5.5 Leimkuhler curve	155
Exercises			156

Bibliography — 159

Index — 175

List of Figures

1.1 Probability mass function of the binomial distribution $\mathcal{B}i(m, p)$ with $m = 5$ and $p = 0.2$ (top left), 0.3 (top right), 0.5 (bottom left) and 0.7 (bottom right)   6

1.2 Probability mass function of the Poisson distribution $\mathcal{P}(\lambda)$ for different values of the parameter. $\lambda = 1$ (top left), 2 (top right), 5 (bottom left) and 7 (bottom right)   8

1.3 Probability mass function of the geometric distribution with parametrization (1.6) and $p = 0.3$ (left) and $p = 0.5$ (right)   15

1.4 Probability density function of the normal distribution (1.7) for different values of the parameters μ and σ 18

1.5 Probability density function of the lognormal distribution (1.8) for different values of the parameter μ and $\sigma = 1$ 19

1.6 Probability density function of the gamma distribution (1.9) with $\sigma = 1.5$ and different values of the parameter α 20

1.7 Probability density function of the exponential distribution (1.12) for different values of the parameter σ 21

1.8 Probability density function of the Weibull distribution (1.13) with $\sigma = 1.5$ and different values of the parameter α 23

1.9 Probability density function of the inverse Gaussian distribution (1.14) with $\lambda = 1.5$ and different values of the parameter μ 24

1.10 Probability density function of the Pareto distribution (1.16) with $\sigma = 1.5$ and different values of the parameter α 26

1.11 Probability density function of the Lomax distribution (1.17) with $\sigma = 1.5$ and different values of the parameter α 27

1.12 Probability density function of the beta distribution (1.18) with $\beta = 0.5$ and different values of the parameter α 29

1.13 Probability density function (left panel) and contour plot (right panel) of the bivariate normal distribution (1.25) with parameter values $\mu_1 = 1$, $\mu_2 = 1.5$, $\sigma_1 = 0.5$, $\sigma_2 = 0.4$ and $\rho = 0.7$ 40

1.14 Q-Q plot. Sample quantiles are obtained from 1000 random variates from lognormal with $\mu = 0.5$ and $\sigma = 1$. Theoretical quantiles are obtained from lognormal with $\hat{\mu} = 0.4909$ and $\hat{\sigma} = 1.0256$ 46

3.1 Observed count of the length of stay variable   92

3.2 Graphs of the pmf given in (3.3) for special cases of parameters α and θ 95

3.3 Observed and expected counts under the model with latent class without covariates ... 98
3.4 Smooth kernel density estimate of the empirical expenditure data ... 101
3.5 Smooth kernel density estimate of the empirical expenditure data and the pdf of the LSN distribution obtained for estimated parameters provided in Table 3.6 ... 102
3.6 Empirical smooth distribution (left) and fitted model (right) ... 114
3.7 Fitted functions for the smoothed variables in the GAM model. From top to down and left to right we have log(EO), log(ED) and log(Age) ... 117
4.1 Marginal distribution in the NHN, NE and NTN models for different values of parameters ... 129
4.2 Zipf plots for the size of the French communes (years 1962, 1975, 1990, 1999, 2006 and 2012) ... 139
4.3 Pareto LC for special values of its parameters: $\alpha = 1.1$ (dashed), $\alpha = 1.5$ (thin) and $\alpha = 2$ (thick) ... 152
4.4 Empirical and fitted Lorenz curves based on 1977 CPS data for cross-sectional family, in the U.S.A ... 154
4.5 Aggarwal LC for special values of its parameters. $\alpha = 0.5$ (dashed), $\alpha = 2$ (thin) and $\alpha = 3$ (thick) ... 155
4.6 Plot of the Leimkuhler curves of the classical Pareto distribution for special values of its parameters. $\alpha = 1.1$ (dashed), $\alpha = 1.5$ (thin) and $\alpha = 2$ (thick) ... 156

List of Tables

2.1	Observed and fitted automobile insurance claims for models in the $(a, b, 0)$ class	58
3.1	Example of the first twenty five observations	88
3.2	Tourism data. Summary statistics for each variable. Filtered database	89
3.3	Observed counts for the variable length of stay	93
3.4	Maximum likelihood estimates and standard error (SE) in parenthesis for the data obtained by using (3.2) without including covariates	97
3.5	Results based on the bimodal distribution. The dependent variable is length of stay	98
3.6	Parameters estimates, their p-values in brackets, maximum of the loglikelihood function, AIC and CAIC for the data expenditure at destination without including covariates	101
3.7	Parameters estimates and p-values in brackets, maximum of the loglikelihood function, AIC and CAIC for the data expenditure at destination without including covariates	107
3.8	Results based on LSN, compound Poisson and compound negative binomial models. Dependent variable, aggregate expenditure at destination	107
3.9	Results based on the Farlie-Gumbel Morgenstern copula without including covariates	113
3.10	Results based on the Farlie-Gumbel Morgenstern copula	115
3.11	Results based on OLS and GAM models. Dependent variable, LS	117
4.1	Production data (Greene, 1980a)	128
4.2	Stochastic production frontier estimates	130
4.3	Estimated technical efficiency	131
4.4	Number of communes and some descriptive statistical measures for the size of the French communes	134
4.5	Parameter estimates obtained by ML estimation for the models considered for the size of the French communes	135
4.6	Values of tail index α and unrestricted mixing weight r for the size of the French communes	136
4.7	NLL (above) and HQIC (below) values evaluated at ML estimates of the models considered for the size of the French communes	136

4.8   Kolmogorov-Smirnov test statistic (KS) and its corresponding p-values   137
      (in brackets) for Pareto, lognormal and CLP distributions for the size
      of the French communes
4.9   100 first observations for ACD model   146
4.10 Maximum likelihood estimates, statistics and misspecification tests of   148
      the different ACD(1,1) models
4.11 Some classical parametric LCs   153
4.12 Data for cross-sectional family, in the U.S.A. (source Ryu and Slottje, 1996)   153
4.13 Results for the parameters estimates and MSE and MAX criteria   154
      based on 1977 CPS data for cross-sectional family, in the U.S.A.
      (source Ryu and Slottje, 1996)

Chapter 1

Basic Statistical Distributions

1.1 Introduction

In this chapter we will introduce the elemental distributions that will be used in the following chapters. Readers interested solely in statistical applications of these model aspects may wish to skip this chapter. We begin this chapter by presenting the most essential discrete univariate distributions supported by nonnegative and positive integers. We will continue discussing the univariate distributions of a continuous nature that will be used in the following chapters. Here we discuss some of the most relevant properties of continuous distributions defined in the real line and support in the positive real numbers.

In the next section, two methods for generating new probability distributions will be introduced. The first of them is based on the mixture of distributions. A mixture distribution is the probability distribution that results from assuming that a random variable is distributed according to some parametrized distribution, with some of the parameters of that distribution is considered to be a random variable. The resulting model, also known as unconditional distribution, is the result of marginalizing or integrating over the latent random variable that represents the parameter of the parametrized distribution or conditional distribution. Next, we examine the recently proposed continuous composite models. These models combine several truncated probability density functions through splicing. In this sense, after partitioning the dataset into several domains, different weighted truncated distributions are assumed for various ranges of the random variable. By using this idea, two different methodologies to generate composite models will be discussed in that section.

In the final part of this chapter, we briefly discuss some useful discrete and continuous multivariate distributions to describe the probabilities for a group of random variables. We firstly provide a detailed treatment of multivariate discrete random variables, emphasizing the Poisson case. The methodology proposes in this chapter can be extended to other discrete probabilistic families. Finally, two continuous multivariate distributions are illustrated in this chapter, the multivariate normal or

Gaussian distribution, or joint normal distribution which is a generalization of the normal distribution to higher dimensions. This model is used to describe any set of correlated real-valued random variables, each of them clustering around its mean. One definition is that a random vector is said to be p-variate normally distributed if every linear combination of its p components has a univariate normal distribution. Properties of this multivariate family are examined. Moreover, the bivariate exponential distribution is also examined in this section. Selection and validation of models concludes the chapter.

1.2 Univariate discrete distributions

In this section, we examine the most essential discrete univariate distributions supported by nonnegative and positive integers.

1.2.1 Bernoulli distribution

The Bernouilli random variable arises in random experiments with two possible outcomes: success or failure. The probability of success is denoted by p where as the probability of failure is $1 - p$, with $0 \leq p \leq 1$. This type of random variable is called the *Bernouilli trial or the Bernouilli experiment*. The Bernouilli random variable is defined below:

Definition 1.1 *The random variable:*

$$N = \{Number\ of\ successes\ in\ a\ Bernouilli\ type\ experiment\},$$

it is called as Bernoulli variable, and it is denoted by $N \sim \mathcal{B}er(p)$.

A Bernouilli variable only takes two values, 0 y 1, and its probability mass function is given by

$$\Pr(N = 1) = p, \ \Pr(N = 0) = 1 - p = q,$$

or simply

$$\Pr(N = n) = p^n(1-p)^{1-n}, \ n = 0,\ 1.$$

Moment of order k about the origin is,

$$E(N^k) = 0^k \cdot \Pr(N=0) + 1^k \cdot \Pr(N=1) = p.$$

The mean and variance are given by

$$E(N) = p,\ var(N) = p(1-p).$$

The latter expressions can be derived via the moment generating function (mgf) by differentiation. The latter function is provided by

$$M_N(t) = E(e^{tN}) = 1 - p + pt,$$

with $t \in \mathbb{R}$.

1.2. UNIVARIATE DISCRETE DISTRIBUTIONS

Example 1.1 *An insurance agent makes phone calls to sell life insurances. The result of a call is classified as a success if he sells the policy, which occurs with a probability of 0.3, and failure if he does not sell it. Model this action as a Bernoulli experiment.*

Solution: The random variable

$$N = \begin{cases} 1 & \text{if he sells the insurance policy,} \\ 0 & \text{otherwise,} \end{cases}$$

is of Bernouilli type, with probability of success $p = 0.2$. The probability mass function is

$$\Pr(N = n) = 0.3^n (1 - 0.3)^{1-n}, \quad n = 0, 1.$$

The mean and variance of N are respectively: $E(N) = p = 0.3$ and $var(N) = p(1-p) = 0.21$. □

1.2.2 Binomial distribution

Let us now suppose that we carry out n Bernouilli type experiments satisfying the following two conditions:

1. The experiment are identical, i.e., the probability of success p is the same in all the trials.

2. The experiment are independent, i.e., the outcome of an experiment does not influence in the outcome of the other experiments.

Under these conditions, the binomial distribution is defined as follows:

Definition 1.2 *Let us consider n independent Bernoulli experiment with probability of success p are carried out. Then, the random variable:*

$$N = \{\text{Number of successes in the } n \text{ trials}\},$$

is named binomial, and the corresponding probability distribution, is called binomial distribution.

It is denoted as $X \sim \mathcal{B}i(m, p)$. The probability mass function is given by

$$\Pr(N = n) = \binom{m}{n} p^n (1-p)^{m-n}, \quad \text{with } n = 0, 1, \ldots, m.$$

The random variable can take values in the set $\{0, 1, \ldots, m\}$. If n successes are obtained, then this implies that $m - n$ failures are achieved. Therefore, since the experiments are independent and identically distributed, the probability of n successes and $m - n$ failures is

$$p^n (1-p)^{m-n}.$$

Now, the number of ways to get n successes (and $m - n$ failures) in m trials is obtained by the formula of the permutations with repetition:

$$PR_{n,m-n}^n = \frac{m!}{n!(m-n)!} = \binom{m}{n},$$

which gives rise to the probability law (1.1). The name of binomial distribution comes from the fact that the probabilities can be obtained from the expansion of Newton's binomial $(p+q)^n = 1$.

Stochastic representation

From the definition of the binomial distribution, it can be deduced that this distribution can be represented as a sum of Bernoulli random variables. If $Z_1, \ldots, Z_m$ are independent and identically distributed random variables $\mathcal{B}er(p)$, the the sum of these random variables $N = Z_1 + \cdots + Z_m$ follows a binomial distribution,

$$N = Z_1 + Z_2 + \cdots + Z_m \sim \mathcal{B}(m, p).$$

Then, by using this result, many properties of the binomial distribution are derived.

1.2.3 Moment and probability generating functions

Let us now consider $N \sim \mathcal{B}(m, p)$. By using the stochastic representation (1.1) and taking into account that the random variables Z_i are independent and identically distributed Bernoulli type random variable, the probability generating function (pgf) is,

$$\begin{aligned} P_N(s) &= E(s^N) = E(s^{Z_1+Z_2+\cdots+Z_m}) = E(s^{Z_1})E(s^{Z_2})\cdots E(s^{Z_m}) \\ &= [P_Z(s)]^m = (1 - p + ps)^m, \text{ with } |s| \leq 1, \end{aligned}$$

since Z_i has pgf given by $P_Z(s) = 1 - p + ps$. The mgf is

$$\begin{aligned} M_N(t) &= \sum_{n=0}^{\infty} e^{tm} \binom{m}{n} p^n (1-p)^{m-n} \\ &= \sum_{n=0}^{\infty} \binom{m}{n} (pe^t)^n (1-p)^{m-n} = (pe^t + 1 - p)^m. \end{aligned} \quad (1.1)$$

The moments of the random variable N can be obtained by differentiating $M_N(t)$ with respect to t. By computing the first and second derivative of $M_N(t)$ with respect to t, and evaluating those expressions at 0, we have

$$\left.\frac{\partial M_N(t)}{\partial t}\right|_{t=0} = \left.\frac{mpe^t M_N(t)}{pe^t + 1 - p}\right|_{t=0} = mp, \quad (1.2)$$

$$\begin{aligned} \left.\frac{\partial^2 M_N(t)}{\partial t^2}\right|_{t=0} &= \left.\frac{m(m-1)(pe^t)^2 M_N(t)}{(pe^t + 1 - p)^2}\right|_{t=0} + \left.\frac{mpe^t M_N(t)}{pe^t + 1 - p}\right|_{t=0} \\ &= mp + m(m-1)p^2, \end{aligned} \quad (1.3)$$

1.2. UNIVARIATE DISCRETE DISTRIBUTIONS

From (1.2) and (1.3), it is simple to derive that $E(N) = mp$ and $var(N) = mp(1-p)$.

Convolutions

The sum of binomial random variables with the same probability of success, say p, is a binomially distributed random variable, where the number of trials is the sum of the number of trials of each individual random variable. The following result is verifed:

Theorem 1.1 *If $N_1, \ldots, N_k$ are independent random variables such that*

$$N_1 \sim \mathcal{B}i(m_1, p), \ldots, X_k \sim \mathcal{B}i(m_k, p),$$

then

$$N_1 + \cdots + N_k \sim \mathcal{B}i(m_1 + \cdots + m_k, p).$$

Proof: By using the pgf of the sum of random variables, we obtain

$$\begin{aligned} P_{N_1+\cdots+N_k}(s) &= E(s^{N_1+N_2+\cdots+N_k}) = E(s^{N_1})E(s^{N_2})\cdots E(s^{N_k}) \\ &= (1-p+ps)^{m_1}(1-p+ps)^{m_2}\cdots(1-p+ps)^{m_k} \\ &= (1-p+ps)^{m_1+\cdots+m_k}. \end{aligned}$$

This expression is the pgf of a random variable $\mathcal{B}(m_1 + \cdots + m_k, p)$ which proves the result. ∎

Other properties

In the following, we state some other essential properties of the binomial distribution. The first one is related to the modal value of the distribution and the second is associated to the probabilities when the complementary of the parameter p is considered.

(i) The mode are the values taken by the binomial random variable satisfying that

$$p(m+1) - 1 \leq \text{Mode} \leq p(m+1).$$

The binomial distribution can be unimodal o bimodal. The latter situation occurs when $p(m+1)$ is integer. Then, the two modal values are $p(m+1)$ and $p(m+1) - 1$.

(ii) If $N_1 \sim \mathcal{B}i(m, p)$, $N_2 \sim \mathcal{B}i(m, 1-p)$, then,

$$\Pr(N_1 = n) = \Pr(N_2 = m - n).$$

Figure 1.1 shows some examples of the pmf of the binomial distribution for $m = 5$ and different values of the parameter p.

A generalization of the binomial distribution, the quasi binomial distribution, can be viewed in Consul (1974) and Consul (1990). Recently, regression analysis of this generalization was studied by Gómez-Déniz et al. (2020).

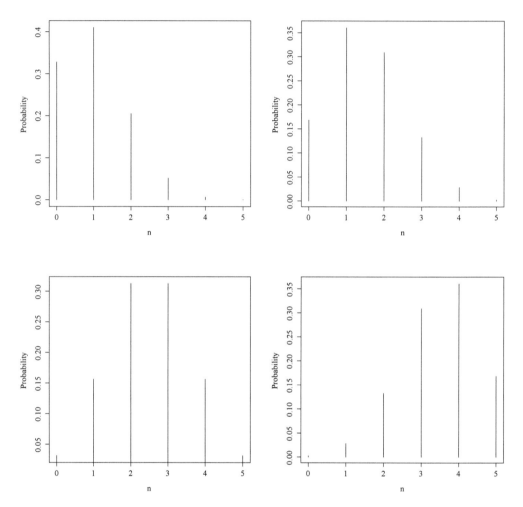

Figure 1.1: Probability mass function of the binomial distribution $\mathcal{B}i(m,p)$ with $m = 5$ and $p = 0.2$ (top left), 0.3 (top right), 0.5 (bottom left) and 0.7 (bottom right).

1.2.4 Poisson distribution

The Poisson distribution naturally arises in many different contexts. For instance, it emerges as the limit of the binomial distribution when the number of trials is large and the probability of success is small. In addition, it can be obtained as the limit of negative binomial and hypergeometric random variables. On the other hand, it is also used as a probabilistic model to describe the occurrence of rare events. Let us suppose that X represents the number of occurrences of an event per unit of time or space. For example:

- Number of annual claims in an insurance portfolio.
- Number of weekly accidents in a particular road section.
- Number of births per hour during a given day.

1.2. UNIVARIATE DISCRETE DISTRIBUTIONS

The experiment consists of counting the number of times that an event occurs in a time or spatial interval of length t. The probability law of the random variable X is determined by the following hypotheses:

(i) The probability of occurrence of a simple event in a small interval of length d is proportional to its duration, that is, $p = cd$, where c is a positive constant.

(ii) Probability of occurrence of two or more events in the same interval is negligible.

(iii) The number of occurrences in an interval is independent of the number of events in another disjoint interval.

(iv) The number of events occurring in two intervals with the same length has the same probability distribution.

By dividing the interval of length t in k subintervals of length $d = t/k$. Then, based on the previous assumptions, we can think of the k subintervals as n independent and identically distributed Bernouilli trials $N_1, \ldots, N_k$ with $\Pr(N_i = 1) = p$, $i = 1, \ldots, k$. Now, $N = \sum_{i=1}^{k} N_i$ follows a binomial distribution with parameters k and $p = ct/k$.

To guarantee that no more than one single event occurs in a subinterval d, the interval can be divided in a large number of subintervals. Therefore, we can obtain the probability mass functions of the random variable N when $k \to \infty$.

As it will be seen later, it is verified that

$$\lim_{k \to \infty} \binom{k}{n} p^k (1-p)^{k-n} = \frac{e^{-ct}(ct)^n}{n!}, \quad n = 0, 1 \ldots$$

By denoting $\lambda = ct$, we have

$$\Pr(N = n) = \frac{e^{-\lambda} \lambda^n}{n!}, \quad n = 0, 1, \ldots; \lambda > 0.$$

Definition 1.3 *We say that a random variable N follows a Poisson distribution and it is represented by $N \sim \mathcal{P}(\lambda)$, if its probability mass function is given by*

$$\Pr(N = n) = \frac{e^{-\lambda} \lambda^n}{n!}, \quad n = 0, 1, \ldots$$

where λ is a positive parameter.

Figure 1.2 shows the probability mass function of the Poisson distribution for different values of the parameter λ. It can be seen that the mode of the distribution moves to the right when the parameter λ increases.

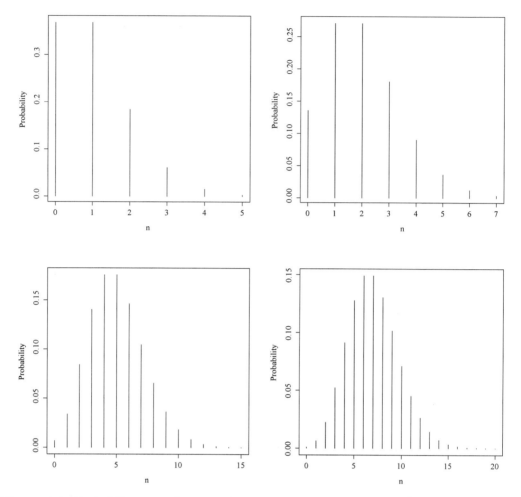

Figure 1.2: Probability mass function of the Poisson distribution $\mathcal{P}(\lambda)$ for different values of the parameter. $\lambda = 1$ (top left), 2 (top right), 5 (bottom left) and 7 (bottom right).

Probability generating and characteristic function

Let us consider a Poisson random variable $N \sim \mathcal{P}(\lambda)$. The pgf of N is obtained as follows

$$\begin{aligned}
P_X(s) = E(s^N) &= \sum_{n=0}^{\infty} s^n \frac{e^{-\lambda}\lambda^n}{n!} \\
&= e^{-\lambda} \sum_{n=0}^{\infty} \frac{(\lambda s)^n}{n!} = e^{\lambda(s-1)}.
\end{aligned}$$

The mgf is

$$M_N(t) = \sum_{n=0}^{\infty} e^{tn} e^{-\lambda} \frac{\lambda^n}{n!} = \exp\{\lambda(e^t - 1)\}, \text{ with } t \in \mathbb{R}. \tag{1.4}$$

1.2. UNIVARIATE DISCRETE DISTRIBUTIONS

The moments of the random variable N can be obtained by differentiating $M_N(t)$ with respect to t. By computing

$$\left.\frac{\partial M_N(t)}{\partial t}\right|_{t=0} = \lambda e^t M_N(t)\big|_{t=0} = \lambda,$$

$$\left.\frac{\partial^2 M_N(t)}{\partial t^2}\right|_{t=0} = \lambda e^t M_N(t) + (\lambda e^t)^2 M_N(t)\big|_{t=0} = \lambda + \lambda^2.$$

The Poisson distribution plays an important role in modelling discrete count data, chiefly due to descriptive adequacy as a model when only randomness is present and the underlying population is homogeneous. The main feature related to the Poisson distribution is equidispersion: its variance is equal to its mean.

From (1.5) and (1.5), it is simple to derive that $E(N) = var(N) = \lambda$. From expression (1.4), the factorial moments can also be easily obtained.

$$E(N(N-1)\cdots(N-k+1)) = P_N^{(k)}(1) = \lambda^k, \quad k = 1, 2, \ldots$$

and therefore

$$\begin{aligned} E(N) &= \lambda, \\ E(N(N-1)) &= \lambda^2, \\ E(N(N-1)(N-2)) &= \lambda^3, \\ E(N(N-1)(N-2)(N-3)) &= \lambda^4. \end{aligned}$$

The skewness and kurtosis coefficients are

$$\begin{aligned} \gamma_1 &= \frac{1}{\sqrt{\lambda}}, \\ \gamma_2 &= \frac{1}{\lambda}. \end{aligned}$$

Therefore, from (1.5) and (1.5), the Poisson distribution is right-skewed and leptokurtic for any value of λ.

Limit of the binomial distribution

As mentioned earlier, if $N \sim \mathcal{B}(m, p)$, m tends to infinity, p to zero and $\lambda = mp$ remains fixed, the binomial distribution converges to the Poisson random variable with parameter $\lambda = mp$:

$$\Pr(N = n) = \binom{m}{n} p^n (1-p)^{m-n} \simeq \frac{e^{-\lambda}\lambda^n}{n!},$$

this approximation is useful if $p < 0.1$ and $mp \leq 10$.

Theorem 1.2 *The limit of the binomial distribution is the Poisson distribution, when m tends to ∞, p tends to zero and the product $\lambda = mp$ remains constant.*

Proof: Since $\lambda = mp$ is constant, then $p = \lambda/m$. Now, by using (1.1) and calculating the limit when m tends to infinity we get,

$$\lim_{m\to\infty} \left(1 - p + pe^t\right)^m = \lim_{m\to\infty} \left[1 + p(e^t - 1)\right]^m$$
$$= \lim_{m\to\infty} \left[1 + \frac{\lambda}{m}(e^t - 1)\right]^m = e^{\lambda(e^t - 1)},$$

which corresponds to (1.4), that is, the mgf of the Poisson distribution with mean $\lambda > 0$. ∎

Convolutions

The Poisson distribution is reproductive with respect to the parameter λ, i.e., the sum of k independent Poisson random variables is a Poisson random variable.

Theorem 1.3 *Let $N_i \sim \mathcal{P}(\lambda_i)$, $i = 1, 2, \ldots, k$ be Poisson random variables. Then the sum random variable $N_1 + N_2 + \cdots + N_k$ is again of Poisson type, i.e.,*

$$N_1 + N_2 + \cdots + N_k \sim \mathcal{P}(\lambda_1 + \lambda_2 + \cdots + \lambda_k).$$

Proof: This can be proved by using the pgf

$$P_{N_1 + N_2 + \cdots + N_k}(s) = P_{N_1}(s) \cdots P_{N_k}(s) = e^{(\lambda_1 + \cdots + \lambda_k)(s-1)}$$

and therefore $N_1 + N_2 + \cdots + N_k \sim \mathcal{P}(\lambda_1 + \cdots + \lambda_k)$. ∎

Example 1.2 *An insurance company has a portfolio that includes 2000 policies. Each of these policies provides coverage for a risk. The compensation for each claim is $3000. Calculate:*

a) Probability of having less than two claims.

b) Probability of having at least 3 claims.

c) Expectation of total compensation claims.

Solution:

a) Let us suppose that the claims occur independently and identically distributed for each one of the policyholders, therefore

$$N = \{\text{Number of claims}\} \sim \mathcal{Bi}(m = 2000, p = 0,001).$$

Since m is large and p small, by using the Poisson approximation

$$N \approx \mathcal{P}(\lambda), \text{ where } \lambda = mp = 2.$$

Then, $\Pr(N < 2) = \Pr(N = 0) + \Pr(N = 1) = e^{-2} + 2e^{-2} = 0.4060.$

1.2. UNIVARIATE DISCRETE DISTRIBUTIONS

b) We can calculate the probability of the complementary event
$$\begin{aligned}\Pr(X \geq 3) &= 1 - \Pr(X < 3) \\ &= 1 - \Pr(X = 0) - \Pr(X = 1) - \Pr(X = 2) = 0.5940.\end{aligned}$$

c) The total compensation C is
$$C = 5000 \cdot N \Longrightarrow E(C) = 5000 \cdot E(N) = \$10000.$$

□

A nice generalization of the Poisson distribution was provided by Consul and Mittal (1973).

1.2.5 Negative binomial distribution

The negative binomial random variable is a generalization of the geometric distribution. Let us now suppose that we carry out independent Bernoulli trials with probability of success p. We now define the random variable

$$N = \{\text{Number of failures before the } r\text{-th success}\}.$$

The probability mass function of (1.5) is provided by

$$\Pr(N = n) = \binom{n+r-1}{n} p^r (1-p)^n, \quad n = 0, 1, 2, \ldots$$

Likewise, for the n failure occurs before the r success, we must run $n + r$ trials, where the last one must be a success. The total number of ways to obtain n failures and $r-1$ successes in $n+r-1$ trials are obtained by the permutation with repetition formula,

$$PR_{n,r-1}^{n+r-1} = \frac{(n+r-1)!}{n!(r-1)!} = \binom{n+r-1}{n}.$$

Finally, by multiplying the latter expression by $(1-p)^n p^r$, we obtain the probability mass function. Now we provide the following definition

Definition 1.4 *A random variable N follows a negative follows a negative binomial distribution if the probability mass function is given by*

$$\Pr(N = n) = \binom{n+r-1}{n} p^r (1-p)^n, \quad n = 0, 1, 2, \ldots,$$

where $0 < p < 1$, $r > 0$, and henceforward it will be represented by $N \sim \mathcal{NB}(r,p)$. An alternative parametrization is

$$\Pr(N = n) = \binom{n+r-1}{n} \left(\frac{1}{1+\beta}\right)^r \left(\frac{\beta}{1+\beta}\right)^n, \quad n = 0, 1, 2, \ldots,$$

where $\beta > 0$ and $r > 0$. In this latter case, we write $N \sim \mathcal{NB}(r, p = 1/(1+\beta))$.

The negative binomial name arises from the fact that the probability can be obtained from the series expansion with negative exponent of $p = 1 - q$,

$$p^{-r} = (1-q)^{-r} = \sum_{n=0}^{\infty} \binom{n+r-1}{n} q^n.$$

The parameter r has not to be necessarily integer, it could be a positive real number, i.e., $r > 0$.

Probability and moment generating functions

From the expression (1.5) it is derived the pgf of (1.5) that is given by

$$P_N(s) = E(s^N) = \frac{p^r}{[1-(1-p)s]^r}, \quad \text{si} \ |s| < 1/(1-p),$$

The mgf is

$$\begin{aligned} M_N(t) &= \sum_{n=0}^{\infty} e^{tn} \binom{n+r-1}{n} p^r (1-p)^n \\ &= \frac{p^r}{(1-(1-p)e^t)^r} \sum_{n=0}^{\infty} \binom{n+r-1}{n} (1-(1-p)e^t)^r (e^t(1-p))^n \\ &= \frac{p^r}{(1-(1-p)e^t)^r}, \end{aligned}$$

given that $0 < (1-p)e^t < 1$, or $t < -log(1-p)$. The moments of the random variable N can be again obtained by differentiating $M_N(t)$ with respect to t at 0.

In particular the mean and variance of a random variable N that follows a negative binomial distribution are $E(N) = rp/(1-p)$ and $var(N) = rp/(1-p)^2$ respectively. Also using (1.5) these quantities are given by

$$\begin{aligned} E(X) &= r\beta, \\ var(X) &= r\beta(1+\beta). \end{aligned}$$

Taking (1.5), we obtain the positive factorial moment of order k:

$$\begin{aligned} E(N(N-1)\cdots(N-k+1)) &= P_N^{(k)}(1) \\ &= r(r+1)\cdots(r+k-1)\left(\frac{1-p}{p}\right)^k, \\ &= \frac{\Gamma(r+k)}{\Gamma(r)} \left(\frac{1-p}{p}\right)^k, \end{aligned}$$

where $\Gamma(\cdot)$ is the Euler (complete) gamma function defined as

$$\Gamma(z) = \int_0^{\infty} u^{z-1} e^{-u} \, du. \tag{1.5}$$

1.2. UNIVARIATE DISCRETE DISTRIBUTIONS

In particular, the the third and fourth factorial moments of the negative binomial distribution are given by

$$E(N(N-1)(N-2)) = \frac{r(r+1)(r+2)(1-p)^3}{p^3},$$

$$E(N(N-1)(N-2)(N-3)) = \frac{r(r+1)(r+2)(r+3)(1-p)^4}{p^4}.$$

From these expressions the skewness and kurtosis coefficients are given by

$$\gamma_1 = \frac{2-p}{\sqrt{r(1-p)}},$$

$$\gamma_2 = \frac{1}{r}\left(6 + \frac{p^2}{1-p}\right),$$

respectively. Therefore the negative binomial distribution is right-skewed and leptokurtic.

Convolutions

If $N_1 \sim \mathcal{NB}(r_1, p), \ldots, N_k \sim \mathcal{NB}(r_k, p)$ are independent random variables, then $N_1 + \cdots + N_k$ is again negative binomial with parameters $r_1 + \cdots + r_k$ and p.

Other properties

Let us consider, for the case $\lambda > 0$, the random variable $N_r \sim \mathcal{NB}(r, p = 1/(1+\beta))$ where $\beta = \lambda/r$. Then, the limit of N_r when r tends to infinity is a random variable of Poisson type. By using (1.5), the pgf of N_r is,

$$P_{N_r}(s) = \frac{1}{[1-\beta(s-1)]^r} = \frac{1}{[1-\frac{\lambda}{r}(s-1)]^r} = \left[1 - \frac{\lambda}{r}(s-1)\right]^{-r}.$$

The limit when r tends to infinity is

$$\lim_{r \to \infty} P_{N_r}(s) = \lim_{r \to \infty} \left[1 - \frac{\lambda}{r}(s-1)\right]^{-r} = e^{\lambda(s-1)},$$

that corresponds to a pgf of a Poisson random variable with mean $\lambda > 0$.

The negative binomial distribution can be also be obtained as a mixture of a Poisson distribution assuming that its mean is a random variable that follows a gamma distribution. See the Example 1.4.

1.2.6 The geometric distribution

This probabilistic model is a special case of the negative binomial distribution when the $r = 1$. The probability mass function of the geometric distribution is given by

$$\Pr(X = n) = p(1-p)^n, \quad n = 0, 1, 2, \ldots \tag{1.6}$$

and $0 < p < 1$.

An alternative parametrization of the geometric distribution is given by

$$\Pr(N = n) = \frac{1}{1+\beta}\left(\frac{\beta}{1+\beta}\right)^n, \quad n = 0, 1, 2, \ldots,$$

where $\beta > 0$. Here, $X \sim \mathcal{G}e(p = 1/(1+\beta))$.

The pgf of (1.7) is

$$P_N(s) = E(s^N) = \sum_{n=0}^{\infty} s^n p(1-p)^n = \frac{p}{1-(1-p)s}, \quad \text{si } |s| < 1/(1-p),$$

In terms of the parametrization given in (1.7), these function is given by

$$P_N(s) = \frac{1}{1 - \beta(s-1)}, \quad |s| < (1+\beta)/\beta.$$

The mgf is

$$M_N(t) = \frac{p}{1 - (1-p)e^t},$$

given that $0 < (1-p)e^t < 1$, or $t < -\log(1-p)$. The moments of the random variable N can be again obtained by differentiating $M_N(t)$ with respect to t at 0.

In particular the mean and variance of a random variable N that follows a geometric distribution are $E(N) = p/(1-p)$ and $var(N) = p/(1-p)^2$ respectively.

For the parametrization (1.7) the mean and variance are

$$\begin{aligned} E(X) &= \beta, \\ var(X) &= \beta(1+\beta). \end{aligned}$$

respectively.

Other properties

The geometric distribution satisfies the memoryless properties. If $X \sim \mathcal{G}e(p)$ with probability function (1.7) it is verified that

$$\Pr(X \geq n + m | X \geq m) = \Pr(X \geq n),$$

for $n, m = 0, 1, 2, \ldots$

Example 1.3 *The probability that an inexperienced insurance agent sells an insurance policy after a phone call is 0.05. Assuming sales between calls are independent with a maximum value of one unit, calculate the probability of selling his first policy in the fifth visit. Obtain the expected number of calls to achieve his first sale.*

1.2. UNIVARIATE DISCRETE DISTRIBUTIONS

Solution: The random variable $X = \{$Number of calls to achieve the first sale$\}$, follows a geometric distribution with parametrization given by (1.6) with parameter $p = 0.05$. Therefore,

$$\Pr(X = 5) = 0.05(1 - 0.05)^{5-1} = 0.0407 = 4.07\%.$$

The expected number of calls to achieve the first sale is given by

$$E(X) = \frac{1}{p} = \frac{1}{0.05} = 20 \text{ calls.}$$

□

Figure 1.3 shows the probability mass function of the geometric distribution defined in (1.6) for different values of the parameter p. It can be seen that the probability at 0 increases with the value of p.

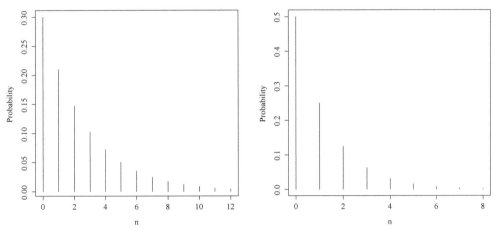

Figure 1.3: Probability mass function of the geometric distribution with parametrization (1.6) and $p = 0.3$ (left) and $p = 0.5$ (right).

Various generalizations of the geometric distribution have been proposed in the statistical literature. See, for example Philippou et al. (1983), Tripathi et al. (1987) and Gómez-Déniz (2010), among others.

1.2.7 Logarithmic distribution

Definition 1.5 *A random variable N follows a logarithmic distribution if its probability mass function is given by,*

$$\Pr(N = n) = -\frac{1}{\log(1 - \theta)} \cdot \frac{\theta^n}{n}, \quad n = 1, 2, \ldots,$$

where $0 < \theta < 1$.

Note that the logarithmic distribution is not defined at 0. The probabilities $\Pr(N = n)$ are always decreasing with respect to n. The pgf is given by,

$$P_N(s) = E(s^N) = \frac{\log(1-\theta s)}{\log(1-\theta)}, \text{ for } s < 1/\theta.$$

whereas the mgf is

$$M_N(t) = E(e^{tN}) = \frac{\log(1-\theta e^t)}{\log(1-\theta)}, \text{ for } t < -\log\theta.$$

The mean and the variance of this distribution are

$$E(X) = \frac{a\theta}{1-\theta},$$
$$var(X) = \frac{a\theta(1-a\theta)}{(1-\theta)^2} = \mu\left(\frac{1}{1-\theta} - \mu\right),$$

where $a = -1/\log(1-\theta)$ and $\mu = E(X)$. These two expressions can be obtained from the mgf. The logarithmic distribution has been used to model the number of individuals of a species.

1.3 Univariate continuous distributions

In this section we will study various continuous parametric probabilistic models, and we will provide their more relevant properties and some important risk quantities.

1.3.1 Normal distribution

The normal distribution plays an important role in statistical modelling. For example, standardized test scores of exams typically resemble a normal or bell shape curve. Also height, athletic ability, and many social and political behaviours of a given population also usually resemble a normal curve. In finance, changes in the log values of certain rates, price indices, and stock prices are assumed to be normally distributed. In actuarial science, the normal distribution is usually used as actuarial statistics. For example, if a large volume loss data is available, the normal approximation is used in the individual and collective risk model (see Section 2.2.2). The probability density function (pdf) of a normal distribution $\mathcal{N}(\mu, \sigma^2)$ is given by,

$$f(x; \mu, \sigma) = \frac{1}{\sigma\sqrt{2\pi}} \exp\left\{-\frac{1}{2}\left(\frac{x-\mu}{\sigma}\right)^2\right\}, \quad -\infty < x < \infty, \quad (1.7)$$

where $\mu \in \mathbb{R}$ y $\sigma > 0$. The distribution function is,

$$F(x; \mu, \sigma) = \Phi\left(\frac{x-\mu}{\sigma}\right),$$

1.3. UNIVARIATE CONTINUOUS DISTRIBUTIONS

where $\Phi(z)$ is the distribution function of the standard normal $\mathcal{N}(0,1)$ given by the expression

$$\Phi(x) = \int_{-\infty}^{x} \frac{1}{\sqrt{2\pi}} \exp\left\{-\frac{1}{2}z^2\right\} dz.$$

An important relationship is that if $X \sim \mathcal{N}(\mu, \sigma^2)$ and if $Z = (X - \mu)/\sigma$, then $Z \sim \mathcal{N}(0,1)$.

The mgf of the normal distribution is given by the expression

$$M_X(t) = \exp\left\{\mu t + \frac{1}{2}\sigma^2 t\right\}.$$

By differentiating (1.8) with respect to t and evaluating the expression at 0, we can derive the central moments of X. In particular we have that $\mathrm{E}(X) = \mu$ and $var(X) = \sigma^2$.

Figure 1.4 show the pdf of the normal distribution for different values of the location parameter μ and scale parameter σ. It can be seen that for the case $\mu = 0$ and $\sigma = 1$, the standard normal distribution is obtained (dashed curve). When $\mu = 0$ and $\sigma = 1$, the mode is located at 0 and the distribution has thicker tails than the standard normal (thin solid curve). For negative values of the location parameter the mode of the distribution moves to the left and values of the scale parameter lower than 1 makes the tail thinner than the standard normal (thick solid curve). Finally, for positive values of the location parameter μ, the mode of the distribution moves to the right. In this case with a fatter tail than the standard normal since σ is larger than 1 (dotted curve).

The skew-normal distribution studied by Azzalini (1985) constitutes the most interesting generalization, among others, of the normal distribution.

1.3.2 Lognormal distribution

The lognormal distribution is one of the most used parametric models in finance and risk theory. It said that a random variable X follows a lognormal distribution if

$$\log X \sim \mathcal{N}(\mu, \sigma^2).$$

It is denoted as $X \sim \mathcal{LN}(\mu, \sigma^2)$. The probability and density functions of the lognormal distribution are obtained by applying a change of variable and they are given by,

$$\begin{aligned} F(x; \mu, \sigma) &= \Phi\left(\frac{\log x - \mu}{\sigma}\right), \quad x > 0, \\ f(x; \mu, \sigma) &= \frac{1}{x\sigma\sqrt{2\pi}} \exp\left\{-\frac{1}{2}\left(\frac{\log x - \mu}{\sigma}\right)^2\right\}, \quad x > 0, \end{aligned} \quad (1.8)$$

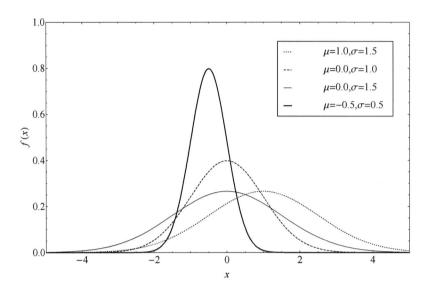

Figure 1.4: Probability density function of the normal distribution (1.7) for different values of the parameters μ and σ.

respectively. Therefore, the probabilities for the lognormal distribution can be easily derived from the standard normal distribution.

Even though the mgf of the lognormal distribution does not exist, the non-central moments of order n are determined by the expression:

$$E(X^n) = E(e^{\log X^n}) = E(e^{n \log X}) = M_{\log X}(n) = \exp\left\{\mu n + \frac{1}{2}\sigma^2 n\right\}.$$

Figure 1.5 shows the pdf of the lognormal distribution for different values of the location parameter μ and a fixed value of the parameter $\sigma = 1$. It can be seen that the lower the value of μ, the thinner is the tail of the distribution.

1.3.3 Gamma distribution

The gamma distribution is one the most used parametric models in risk analysis when unimodal, right-skewed, and positive data are considered. This distribution is also used in finance to price options on assets as an alternative to the lognormal model when the distribution of the assets is positively skewed. A random variable X follows a gamma distribution with parameters α and σ (represented by $X \sim \mathcal{G}(\alpha, \sigma)$) if its pdf can be written as

$$f(x; \alpha, \sigma) = \frac{x^{\alpha-1} \exp(-x/\sigma)}{\sigma^\alpha \Gamma(\alpha)}, \quad \text{for } x > 0 \tag{1.9}$$

where α, σ are positive real numbers, and α is a shape parameter and σ is an scale parameter. Note that by taking $\beta = 1/\sigma$, we have that β is the rate parameter of

1.3. UNIVARIATE CONTINUOUS DISTRIBUTIONS

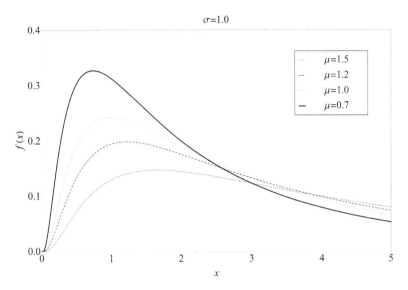

Figure 1.5: Probability density function of the lognormal distribution (1.8) for different values of the parameter μ and $\sigma = 1$.

the gamma distribution. The cumulative distribution function (cdf) of (1.9) can be written as,

$$F(x; \alpha, \sigma) = \Gamma(x/\sigma, \alpha), \quad x > 0,$$

where $\Gamma(X, a)$ is the incomplete gamma function that is defined as

$$\Gamma(x, a) = \frac{1}{\Gamma(a)} \int_0^x t^{a-1} \exp(-t)\, dt. \quad x > 0, \qquad (1.10)$$

The gamma distribution includes as particular cases the Erlang $\alpha \in \mathbb{N}$), in this case the cdf is given by

$$F(x; \alpha, \sigma) = 1 - \sum_{k=0}^{\alpha-1} \exp(-x/\sigma) \frac{x^k}{\sigma^k k!}, \quad x \geq 0.$$

The mgf is given by

$$\begin{aligned} M_X(t) &= \int_0^\infty e^{tx} \frac{x^{\alpha-1} e^{-\frac{x}{\sigma}}}{\sigma^\alpha \Gamma(\alpha)}\, dx = \frac{1}{\sigma^\alpha \Gamma(\alpha)} \int_0^\infty x^{\alpha-1} e^{-x\left(\frac{1}{\sigma}-t\right)}\, dx \\ &= \frac{1}{\sigma^\alpha \Gamma(\alpha)} \frac{\Gamma(\alpha)}{(1/\sigma - t)^\alpha} = (1 - \sigma t)^{-\alpha}, \text{ for } t < 1/\sigma. \end{aligned} \qquad (1.11)$$

Differentiating (1.11) with respect to t, the non-central moments are given by

$$E(X^n) = \frac{\Gamma(\alpha + n) \sigma^n}{\Gamma(\alpha)}.$$

In particular we have that the mean and variance of the gamma distribution are $E(X) = \alpha\sigma$ and $var(x) = \alpha\sigma^2$. By using the alternative parametrization these quantities are $E(X) = \alpha/\beta$ and $var(x) = \alpha/\beta^2$ respectively.

The exponential distribution is obtained when ($\alpha = 1$), the chi-squared distribution is a particular case of the gamma distribution when $\sigma = 2$ and $\alpha = \nu/2$ where $\nu \in \mathbb{N}$ is the degrees of freedom, and the normal is obtained as a limiting case if $\alpha \to \infty$.

Figure 1.6 show the pdf of the gamma distribution for different values of the scale parameter α and a fixed value of the scale parameter $\sigma = 1.5$. It can be observed that the larger the value of α, the thicker is the tail of the distribution.

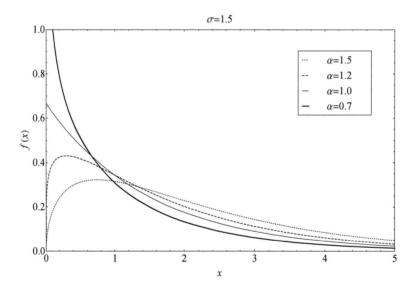

Figure 1.6: Probability density function of the gamma distribution (1.9) with $\sigma = 1.5$ and different values of the parameter α.

1.3.4 Exponential distribution

The exponential distribution is a particular cases of the gamma distribution when $\alpha = 1$. It is defined in terms of its pdf,

$$f(x;\sigma) = \frac{1}{\sigma}\exp(-x/\sigma), \quad x > 0, \tag{1.12}$$

where $\sigma > 0$ is the scale parameter. By taking $\beta = 1/\sigma$ the exponential distribution is specified in terms of the rate parameter. The cdf is

$$F(x) = 1 - \exp(-x/\sigma), \quad x \geq 0.$$

It is used to model processes with a constant failure rate and it is characterized as being the only continuous distribution that verifies the memoryless property. This means that when the random variable X is interpreted as the waiting time for an

1.3. UNIVARIATE CONTINUOUS DISTRIBUTIONS

event to occur relative to some initial time t, this relation implies that, if X is conditioned on a failure to observe the event over some initial period, the distribution of the remaining waiting time is the same as the original unconditional distribution:

$$\Pr(X > x + t | X > t) = \Pr(X > x), \text{ with } x, t \geq 0.$$

The mgf is given by

$$\begin{aligned} M_X(t) &= \int_0^\infty e^{tx} \frac{e^{-\frac{x}{\sigma}}}{\sigma} dx = \frac{1}{\sigma} \int_0^\infty e^{-x(\frac{1}{\sigma} - t)} dx \\ &= \frac{1}{\sigma} \frac{1}{(1/\sigma - t)} = (1 - \sigma t)^{-1}, \text{ for } t < 1/\sigma. \end{aligned}$$

Differentiating (1.13) with respect to t, the non-central moments are given by

$$E(X^n) = \Gamma(1 + n)\sigma^n.$$

In particular we have that the mean and variance of the exponential distribution are $E(X) = \sigma$ and $var(x) = \sigma^2$. By using the alternative parametrization these quantities are $E(X) = 1/\beta$ and $var(x) = 1/\beta^2$ respectively.

Figure 1.7 shows the pdf of the exponential distribution for different values of the scale parameter σ. It can be seen that the larger is value of σ, the thicker is the tail of the distribution.

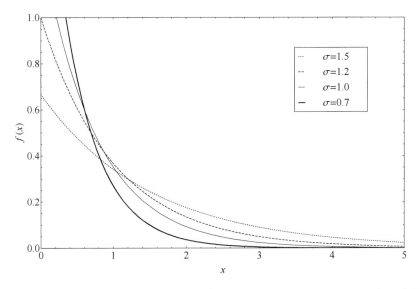

Figure 1.7: Probability density function of the exponential distribution (1.12) for different values of the parameter σ.

1.3.5 Weibull distribution

Reliability engineering may be described as the use of applied statistics for engineering evaluation purposes. Naturally, the statistical techniques incorporated in reliability engineering can be used for many other purposes. The Weibull distribution has been used during many years in reliability engineering. Over the last decades, it has been used in financial and risk analysis. For example some of these applications include credit card default, life insurance, and stock brokerage. The Weibull distribution generalizes the exponential distribution, and it admits constant, positive, and negative failure rates. This distribution is also used as minimum extreme value distribution. In this regard, under certain restrictions, the minimum of distributions without lower bounds tend to the smallest extreme value distribution and the minimum of distributions bounded below tends to the Weibull.

It can be defined in terms of the cdf,

$$F(x;\alpha,\sigma) = 1 - \exp\left[-\left(\frac{x}{\sigma}\right)^\alpha\right], \quad x \geq 0,$$

where $\sigma > 0$ is a scale parameter, and $\alpha > 0$ is a shape parameter. If $\alpha = 1$ the exponential distribution is obtained and if $\alpha = 2$ and $\mu = 0$ the *Rayleigh* distribution is achieved.

The pdf is given by

$$f(x;\alpha,\sigma) = \frac{\alpha}{\sigma}\left(\frac{x}{\sigma}\right)^{\alpha-1} \exp\left[-\left(\frac{x}{\sigma}\right)^\alpha\right], \quad x > 0. \tag{1.13}$$

The failure or hazard rate function is provided by

$$h(x;\alpha,\sigma) = \frac{\alpha}{\sigma}\left(\frac{x}{\sigma}\right)^{\alpha-1},$$

the hazard is decreasing for $\alpha < 1$, increasing for $\alpha > 1$ and constant for $\alpha = 1$, in the latter case, the Weibull distribution reduces to an exponential distribution. The mgf of the Weibull distribution is given by

$$\begin{aligned} M_X(t) &= \int_0^\infty e^{tx} \frac{\alpha}{\sigma}\left(\frac{x}{\sigma}\right)^{\alpha-1} \exp\left[-\left(\frac{x}{\sigma}\right)^\alpha\right] dx \\ &= \sum_{n=0}^\infty \frac{t^n \sigma^n}{n!} \Gamma\left(1 + \frac{n}{\alpha}\right). \end{aligned}$$

From this expression and differentiating with respect to t, the mean and variance of the Weibull distribution are

$$E(X) = \sigma \Gamma\left(1 + \frac{1}{\alpha}\right),$$

$$var(X) = \sigma^2 \left\{\Gamma\left(1 + \frac{2}{\alpha}\right) - \Gamma\left(1 + \frac{1}{\alpha}\right)^2\right\}.$$

Figure 1.8 shows the pdf of the Weibull distribution for different values of the shape parameter α and fixed value of the scale parameter $\sigma = 1.5$. It is observed that for values lower than 1 for α, the mode of the distribution is located at 0.

The generalization of the gamma distribution proposed by Stacy (1962) includes as special cases the gamma, exponential and Weibull distribution.

1.3. UNIVARIATE CONTINUOUS DISTRIBUTIONS

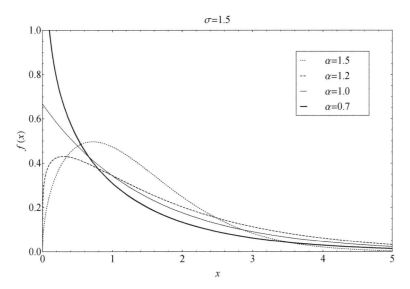

Figure 1.8: Probability density function of the Weibull distribution (1.13) with $\sigma = 1.5$ and different values of the parameter α.

1.3.6 Inverse Gaussian distribution

Specifically, we consider a distribution that is widely used in studies of frailty, the inverse Gaussian distribution (Hougaard, 1984). The inverse Gaussian or Wald distribution has many applications in studies of life time, reaction times, reliability and number of event occurrences Tweedie (1957); (Lancaster, 1972; Chhikara and Folks, 1977; Jørgensen, 1982; Seshadri, 1983; Chhikara and Folks, 1989 and Balakrishnan and Nevzorov, 2003; among others), and it has been applied in fields such as economics, agriculture, demography, ecology, engineering, genetics, meteorology, and the internet (Seshadri, 1999), in the study of many different topics, including financial asset returns, turbulent wind speeds, impulsive noise in radar, and radar and communication channels. It is a member of the natural exponential family of distributions and can be considered an alternative to exponential, lognormal, log-logistic, Frechet and Weibull distributions, among others. The inverse Gaussian distribution, for example, is as suitable as the gamma, both analytically and computationally, although it is not as widely used. Furthermore, the inverse Gaussian is a less complex alternative to the classical log-Normal model, and its hazard rate function has a ∩-shape like the lognormal, generalized Weibull and log-logistic distributions, i.e., the inverse Gaussian is unimodal, which increases from 0 to its maximum value and then decreases asymptotically to a constant. It is also likely to prove useful in statistical applications as a flexible and tractable model for fitting duration data and other right-skewed unimodal data. Finally, it is a flexible closed-form distribution that can be applied to model heavy-tailed processes.

This distribution was originally used in finance to describe the time a stock reaches a certain price for the first time. It has also been used to model stock returns and interest rate processes. It has also been used in risk analysis and actuarial statistics as an alternative to lognormal and gamma distributions in the context of generalized linear model for the claims size distribution.

This distribution has similar properties to the normal distribution. It is defined in terms of its density function, as follows,

$$f(x; \lambda, \mu) = \sqrt{\frac{\lambda}{2\pi x^3}} \exp\left[-\frac{\lambda(x-\mu)^2}{2\mu^2 x}\right], \quad x > 0, \qquad (1.14)$$

where $\lambda, \mu > 0$ are shape and location parameter respectively. The distribution function is provided by

$$F(x; \lambda, \mu) = \Phi\left(\sqrt{\frac{\lambda}{x}}\left(\frac{x}{\mu} - 1\right)\right) + e^{2\lambda/\mu} \Phi\left(-\sqrt{\frac{\lambda}{x}}\left(\frac{x}{\mu} + 1\right)\right), \quad x > 0,$$

where $\Phi(\cdot)$ is the cdf of the standard normal distribution.

The mgf of the inverse Gaussian distribution is

$$M_X(t) = \exp\left\{\frac{\lambda}{\mu}\left(1 - \sqrt{1 - \frac{2\mu^2 t}{\lambda}}\right)\right\}.$$

From this expression and differentiating with respect to t, the mean and variance of the inverse Gaussian distribution are $E(X) = \mu$ and $var(X) = \mu^3/\lambda$. This distribution is unimodal and positively skewed.

Figure 1.9 shows the pdf of the inverse Gaussian distribution for different values of the location parameter μ and fixed value of the shape parameter $\lambda = 1.5$. It is noticed that the larger is the value of μ, the thicker is the tail of the distribution.

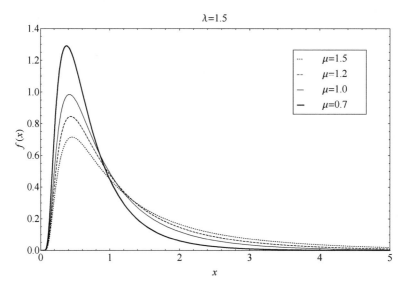

Figure 1.9: Probability density function of the inverse Gaussian distribution (1.14) with $\lambda = 1.5$ and different values of the parameter μ.

1.3. UNIVARIATE CONTINUOUS DISTRIBUTIONS

The inverse Gaussian distribution is a special case of a more general distribution, the generalized inverse Gaussian distribution, with pdf given by,

$$f(x) = \frac{x^{p-1}\mu^{-p}}{2K_p(\lambda/\mu)} \exp\left[-\lambda\left(1 + \frac{x^2}{\mu^2}\right)\right], \quad x > 0, \tag{1.15}$$

where $p \in \mathbb{R}$, $\mu, \lambda > 0$ and $K_\nu(u)$ is the modified Bessel function of third kind with index ν, given by,

$$K_\nu(u) = \frac{1}{2}\int_0^\infty x^{\nu-1} \exp\left[-\frac{u}{2}\left(x + \frac{1}{x}\right)\right] dx.$$

When $p = -1/2$ expression (1.15) reduces to (1.14).

1.3.7 Family of Pareto distributions

The Pareto Distribution was named after the Italian economist and sociologist Vilfredo Pareto. The Pareto Distribution has been used in describing social, scientific, and geophysical phenomena in society. Pareto distributions have been widely used in the actuarial and risk literature for its good properties. The text Arnold (1983) is an excellent reference for studying the Pareto distributions. From the risk analysis point of view, the fit of the tail of a distribution is a key aspect. In this regard, it is important that the events unlikely but with great cost or severity have a probability not negligible. The pdf of the Pareto distribution converges to zero much slower than those of the normal and lognormal distribution. Therefore, it is much safer to use it to calculate premiums for reinsurance in large claims tranches.

It is well known that there are several Pareto type distributions. One of the most used is the classic distribution of Pareto, which is part of the hierarchy of distributions proposed by Arnold (1983). In that work, beginning with the simpler distribution, the distributions are generalized, until a total of four nested models are obtained. In the following, we will study two members of the Pareto family of probability distributions.

1.3.8 Classical Pareto distribution

A random variable X follows a classical Pareto distribution if its pdf is given by,

$$f(x; \alpha, \sigma) = \frac{\alpha\sigma^\alpha}{x^{\alpha+1}}, \quad x > \sigma > 0, \tag{1.16}$$

with $\alpha > 0$ is a shape parameter and σ is an scale parameter, while $f(x) = 0$ if $x < \sigma$.

The cdf is given by

$$F(x; \alpha, \sigma) = 1 - \left(\frac{\sigma}{x}\right)^\alpha, \quad x \geq \sigma > 0,$$

The mgf of the classical Pareto distribution does not exist, however their mean and variance are given by

$$E(X) = \frac{\alpha\sigma}{\alpha-1}, \text{ for } \alpha > 1,$$

$$var(X) = \frac{\alpha\sigma^2}{(\alpha-1)^2(\alpha-2)}, \text{ for } \alpha > 2.$$

Figure 1.10 shows the pdf of the Pareto distribution for different values of the shape parameter α and fixed value of the scale parameter $\sigma = 1.5$. It is observed that the lower the value of α, the thicker is the tail of the distribution.

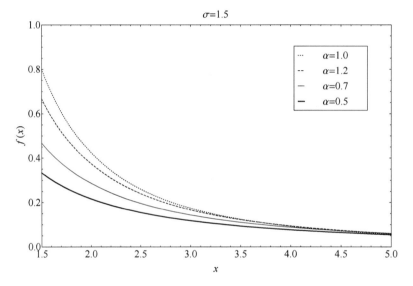

Figure 1.10: Probability density function of the Pareto distribution (1.16) with $\sigma = 1.5$ and different values of the parameter α.

The Pareto distribution is related to the exponential distribution in the following way:

Result 1.1 *The random variable $Y = \log(X/\sigma)$ is exponentially distributed with rate parameter α.*

Proof: By calculating the cdf of Y, we have

$$\Pr(Y < y) = \Pr\left(\log\frac{X}{\sigma} < y\right) = \Pr(X < \sigma e^y)$$

$$= 1 - \left(\frac{\sigma}{\sigma e^y}\right) = 1 - e^{-\alpha y}.$$

1.3. UNIVARIATE CONTINUOUS DISTRIBUTIONS

1.3.9 Pareto type II or Lomax distribution

A random variable X follows a Pareto type II (also known as Lomax or generalized Pareto distribution) if its pdf is given by,

$$f(x; \alpha, \sigma) = \frac{\alpha/\sigma}{(1 + x/\sigma)^{\alpha+1}}, \quad x > 0 \qquad (1.17)$$

and $f(x; \alpha, \sigma) = 0$ if $x < 0$, where $\alpha > 0$ is a shape parameter and σ a scale parameter.

The cdf is given by

$$F(x; \alpha, \sigma) = 1 - \left(\frac{1}{1 + x/\sigma}\right)^{\alpha}, \quad x \geq 0.$$

The mgf of the Pareto type II distribution does not exist, however their mean and variance are given by

$$\begin{aligned} E(X) &= \frac{\sigma}{\alpha - 1}, \text{ for } \alpha > 1, \\ var(X) &= \frac{\alpha \sigma^2}{(\alpha - 1)^2 (\alpha - 2)}, \text{ for } \alpha > 2. \end{aligned}$$

Figure 1.11 shows the pdf of the Pareto type II distribution for different values of the shape parameter α and fixed value of the scale parameter $\sigma = 1.5$. It is noted that the lower the value of α, the thicker is the tail of the distribution.

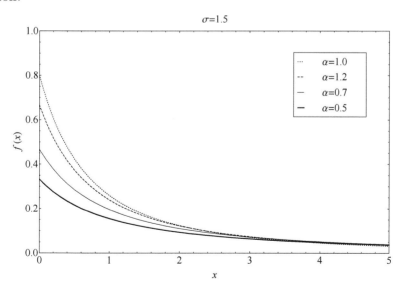

Figure 1.11: Probability density function of the Lomax distribution (1.17) with $\sigma = 1.5$ and different values of the parameter α.

Some generalizations of the Pareto distribution have been proposed in the statistical literature. These can be viewed in Sarabia and Prieto (2009) and Gómez-Déniz and Calderín-Ojeda (2015a). Furthermore, the Kumaraswamy's distribution is a valid alternative to the Beta distribution for studying data with bounded support (see Jones, 2009).

1.3.10 Beta distribution

The beta distribution is a continuous family of probability distributions defined in the interval $(0,1)$. The model distribution can be easily extended to describe random variables that are limited to intervals of finite lengths in a huge variety of fields. The beta distribution is parameterized in terms of two shape parameters $\alpha > 0$ and $\beta > 0$. This distribution is sometimes used to simulate random recovery rates when assessing credit risks.

The pdf of the beta distribution is given by

$$f(x;\alpha,\beta) = \frac{\Gamma(\alpha+\beta)}{\Gamma(\alpha)\Gamma(\beta)} x^{\alpha-1}(1-x)^{\beta-1} \tag{1.18}$$

where $\Gamma(\cdot)$ is the complete gamma function given in (1.5).

In the following for a random variable X that follows a beta distribution will be denoted as $X \sim \mathcal{B}e(\alpha,\beta)$. The cdf of the beta distribution is provided by

$$F(x;\alpha,\beta) = \frac{B(x;\alpha,\beta)}{B(\alpha,\beta)},$$

where $B(\cdot;\cdot,\cdot)$ and $B(\cdot,\cdot)$ are the incomplete and complete beta functions respectively given by

$$\begin{aligned} B(z;a,b) &= \int_0^z u^{a-1}(1-u)^{b-1}\,du, \\ B(a,b) &= \frac{\Gamma(a)\Gamma(b)}{\Gamma(a+b)}. \end{aligned}$$

The mgf of the beta distribution is given by

$$M_X(t) = {}_1F_1(\alpha;\alpha+\beta;t),$$

where ${}_1F_1$ is the Kummer confluent hypergeometric function defined as

$${}_1F_1(a;b;z) = \frac{\Gamma(b)}{\Gamma(a)\Gamma(b-a)} \int_0^1 t^{a-1}(1-t)^{b-a-1} \exp(zt)\,dt. \tag{1.19}$$

The mean of the beta distribution is $E(X) = \frac{\alpha}{\alpha+\beta}$ and the variance is $var(X) = \frac{\alpha\beta}{(\alpha+\beta)^2(\alpha+\beta+1)}$.

Figure 1.12 shows the pdf of the beta distribution for different values of the shape parameter α and $\beta = 0.5$.

1.4. DERIVING NEW DISTRIBUTIONS

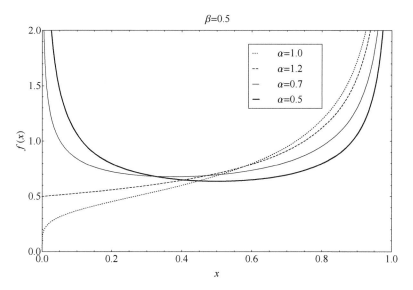

Figure 1.12: Probability density function of the beta distribution (1.18) with $\beta = 0.5$ and different values of the parameter α.

In the statistical literature there is a wide catalog of generalizations of the beta distribution. Without wanting to be exhaustive, we quote the following: Chen and Novick (1984), McDonald (1984), Nadarajah (2005) and Gómez-Déniz and Sarabia (2018).

1.4 Deriving new distributions

In this section we derive new parametric models from existing ones. Different methodologies are discussed in this Section.

1.4.1 Mixture of distribution

We start by defining the density function of the unconditional distribution.

Definition 1.6 *Let X be a random variable with the density function given by $f(x|\theta)$ where θ is a parameter. Let θ be a realization of the random variable Θ with density function given by $f_\Theta(\theta)$. Then the unconditional density of X is*

$$f_X(x) = \int_\Theta f(x|\theta) f_\Theta(\theta) \, d\theta.$$

The resulting distribution is a mixture distribution.

The mean of the resulting distribution can be obtained by the law of iterated expectations:

$$E(X) = E_\Theta(E(X|\theta)).$$

The variance is given by

$$var(X) = E_\Theta(var(X|\theta)) + var_\Theta(E(X|\theta)). \quad (1.20)$$

Example 1.4 *Let us suppose that the number of claims X, follows a Poisson distribution with mean λ. Let us also suppose that this parameter λ follows a gamma distribution with shape parameter $\alpha > 0$ and rate parameter $\beta > 0$. Derive the unconditional distribution of X.*

Solution: The unconditional distribution of X is

$$\begin{aligned}
f(x) &= \int_0^\infty \Pr(X = x|\lambda)\, f(\lambda|\alpha,\beta)\, d\lambda \\
&= \int_0^\infty \frac{e^{-\lambda}\lambda^x}{x!}\, \frac{\beta^\alpha}{\Gamma(\alpha)} \lambda^{\alpha-1} e^{-\beta\lambda}\, d\lambda \\
&= \frac{\beta^\alpha}{x!\Gamma(\alpha)} \int_0^\infty \lambda^{x+\alpha-1} e^{-(\beta+1)\lambda}\, d\lambda \\
&= \frac{\beta^\alpha}{\Gamma(x+1)\Gamma(\alpha)}\, \frac{\Gamma(x+\alpha)}{(\beta+1)^{x+\alpha}} \\
&= \binom{x+\alpha-1}{x} \left(\frac{1}{\beta+1}\right)^x \left(\frac{\beta}{\beta+1}\right)^\alpha.
\end{aligned}$$

This is a negative binomial distribution with parameters α and success probability $p = 1/(\beta+1)$. □

Example 1.5 *Let us suppose that the number of claims X, follows a Poisson distribution with mean λ. Let us also suppose that this parameter λ follows an inverse Gaussian density function given by*

$$f(\lambda;\gamma,\mu) = \sqrt{\frac{\gamma}{2\pi\lambda^3}} \exp\left\{-\frac{\gamma(\lambda-\mu)^2}{2\mu^2\lambda}\right\}, \quad \lambda > 0,$$

with $\gamma, \mu > 0$ are shape and location parameter respectively. Derive the unconditional distribution of X. Also find the mean and variance of this distribution.

Solution: The unconditional distribution of X is

$$\begin{aligned}
f(x) &= \int_0^\infty \Pr(X = x|\lambda)\, f(\lambda|\gamma,\mu)\, d\lambda \\
&= \int_0^\infty \frac{e^{-\lambda}\lambda^x}{x!} \sqrt{\frac{\gamma}{2\pi\lambda^3}} \exp\left\{-\frac{\gamma(\lambda-\mu)^2}{2\mu^2\lambda}\right\} d\lambda \\
&= \sqrt{\frac{\gamma}{2\pi}}\, \frac{1}{x!} \int_0^\infty \lambda^{x-3/2} \exp\left\{-\frac{\gamma(\lambda-\mu)^2}{2\mu^2\lambda} - \lambda\right\} d\lambda \\
&= \frac{\sqrt{\gamma}}{\pi x!} e^{\gamma/\mu} \gamma^{\frac{x}{2}-\frac{1}{4}} \left(\frac{\gamma}{\mu^2}+2\right)^{\frac{1}{4}(1-2x)} K_{x-\frac{1}{2}}\left(\sqrt{\gamma}\sqrt{\frac{\gamma}{\mu^2}+2}\right),
\end{aligned}$$

1.4. DERIVING NEW DISTRIBUTIONS

where $K_z(\cdot)$ is the modified Bessel function of the third kind. This distribution is denoted as Poisson-inverse Gaussian distribution (see for example, Willmot, 1987 and Gómez-Déniz and Calderín-Ojeda, 2018, among others).

By using (1.20), we have

$$E(X) = E_\lambda(E(X|\lambda)) = E_\lambda(\lambda) = \mu.$$

From this result, it is simple to derive a regression model from this distribution denoted as Poisson-inverse Gaussian regression model (see Dean et al., 1989). Similarly, the variance is obtained from (1.20)

$$var(X) = E_\lambda(var(X|\lambda)) + var_\lambda(E(X|\lambda)) = \mu\left(1 + \frac{\mu^2}{\gamma}\right).$$

□

Example 1.6 *We say that a random variable X has a negative binomial-inverse Gaussian distribution if it admits the stochastic representation:*

$$X|\lambda \sim \mathcal{NB}(r, p = e^{-\lambda}), \quad (1.21)$$
$$\lambda \sim \mathcal{IG}(\mu, \gamma), \quad (1.22)$$

with $r, \mu, \psi > 0$. We will denote this distribution by $X \sim \mathcal{NBIG}(r, \mu, \gamma)$ (see Gómez-Déniz et al., 2008).

Let $X \sim \mathcal{NBIG}(r, \mu, \gamma)$ be a negative binomial-inverse Gaussian distribution defined in (1.21)–(1.22). Some basic properties are:

(a) The probability mass function is given by

$$\Pr(X = x) = \binom{r + x - 1}{x} \sum_{j=0}^{x} (-1)^j \binom{x}{j} \exp\left\{\frac{\gamma}{\mu}\left[1 - \sqrt{1 + \frac{2(r+j)\mu^2}{\gamma}}\right]\right\}, \quad (1.23)$$

with $x = 0, 1, 2, \ldots$ and $r, \mu, \gamma > 0$.

(b) The mean, second order moment and variance are given by,

$$E(X) = r[M_\lambda(1) - 1],$$
$$var(X) = (r + r^2)M_\lambda(2) - rM_\lambda(1) - r^2 M_\lambda^2(1),$$

where $M_\lambda(\cdot)$ is defined in (1.15).

Obviously, if $X|\lambda \sim \mathcal{NB}(r, e^{-\lambda})$ and $\lambda \sim \mathcal{IG}(\mu, \gamma)$, the probability mass function of X can be obtained by using the compound formula,

$$\Pr(X = x) = \int_0^\infty \Pr(X = x|\lambda) f(\lambda; \mu, \gamma) \, d\lambda,$$

where $f(\lambda;\mu,\gamma)$ is the pdf of an inverse Gaussian distribution defined in (1.5). Then, making use of the Newton binomial expansion we have that,

$$\Pr(X = x) = \binom{r+x-1}{x} \sum_{j=0}^{x}(-1)^j \binom{x}{j} \int_0^\infty e^{-\lambda j} f(\lambda;\mu,\psi)\,d\lambda,$$

and the result is obtained by using the expression (1.15). Once again the mean and variance of the negative binomial-inverse Gaussian distribution can be obtained via (1.20) and (1.20).

Example 1.7 *Let us suppose that the number of claims X, follows an exponential distribution with mean λ. Let us also suppose that this parameter λ follows an inverse Gaussian density function given by (1.5). Derive the unconditional distribution of X. Also find the mean and variance of this distribution.*

Solution: The unconditional distribution of X is

$$\begin{aligned}
f(x) &= \int_0^\infty f(x|\lambda)\,f(\lambda|\gamma,\mu)\,d\lambda \\
&= \int_0^\infty \frac{1}{\lambda} e^{-\frac{x}{\lambda}} \sqrt{\frac{\gamma}{2\pi\lambda^3}} \exp\left\{-\frac{\gamma(\lambda-\mu)^2}{2\mu^2\lambda}\right\} d\lambda \\
&= \sqrt{\frac{\gamma}{2\pi}} \int_0^\infty \lambda^{-5/2} \exp\left\{-\frac{\gamma(\lambda-\mu)^2}{2\mu^2\lambda} - \frac{x}{\lambda}\right\} d\lambda \\
&= \frac{\sqrt{\gamma}\, e^{\frac{\gamma}{\mu} - \sqrt{\frac{\gamma}{\mu^2}}\sqrt{\gamma+2x}} \left(\sqrt{\frac{\gamma}{\mu^2}} + \frac{1}{\sqrt{\gamma+2x}}\right)}{\gamma + 2x}.
\end{aligned}$$

This distribution is denoted as an exponential-inverse Gaussian distribution. This distribution is a particular case of the gamma-inverse Gaussian distribution in Gómez-Déniz et al. (2013). A regression model based on this distribution was examined by Frangos and Karlis (2004).

By using (1.20), we have

$$E(X) = E_\lambda(E(X|\lambda)) = E_\lambda(\lambda) = \mu.$$

From this result, it is simple to derive a regression model for this distribution. Similarly, the variance is obtained from (1.20)

$$\begin{aligned}
var(X) &= E_\lambda(var(X|\lambda)) + var_\lambda(E(X|\lambda)) \\
&= E_\lambda(\lambda^2) + var_\lambda(\lambda) = \mu^2 + 2\frac{\mu^3}{\gamma}.
\end{aligned}$$

□

1.4.2 Composite models

Composite models combine various truncated densities through splicing. In this regard, after partitioning the data into several domains, different weighted truncated distributions are assumed for various ranges of the random variable. There are several alternatives to define a composite model. In this paper we firstly consider the one used by Scollnik (2007), which improved the model proposed by Cooray and Ananda (2005) incorporating unrestricted mixing weights and assuming continuity and differentiability conditions at the threshold. The density function of the composite model can be written as

$$f(x) = \begin{cases} r\, f_1^*(x), & 0 < x \leq \theta, \\ (1-r)\, f_2^*(x), & \theta < x < \infty, \end{cases}$$

with $0 \leq r \leq 1$, $f_1^*(x) = \frac{f_1(x)}{F_1(\theta)}$ and $f_2^*(x) = \frac{f_2(x)}{1-F_2(\theta)}$ are adequate truncations of the pdf's f_1 and f_2 up to and thereafter of an unknown threshold value θ, where $F_1(\theta)$ and $F_2(\theta)$ denote the cdf of f_1 and f_2 at θ respectively. Then (1.24) can be seen as a convex sum of two density functions and hence it is in a form of a mixture model. After imposing the continuity condition (i.e., $f(\theta^-) = f(\theta^+)$), we have

$$r = \frac{f_2(\theta)\, F_1(\theta)}{f_2(\theta)\, F_1(\theta) + f_1(\theta)\, (1 - F_2(\theta))}.$$

Next, a differentiability condition at θ was also imposed in order to make (1.24) smooth and to reduce the number of parameters.

Alternatively, a more recent approach to define composite models was proposed by Calderín-Ojeda and Kwok (2016) based on a mode-matching process (see also Calderín-Ojeda, 2015 and Calderín-Ojeda et al., 2016, among others). This methodology ensures the continuity and differentiability conditions aforementioned and it also allows the use of any density whose mode can be expressed in closed-form, what facilitates the implementation of the model. The key idea behind this approach consists of using as first component of the continuous composite model, adequate truncation of a chosen distribution (say $f_1^*(x)$) up to the modal value, and the second part of the distribution uses adequate truncation of a second distribution (say $f_2^*(x)$).

The cdf $F_X(x)$ is given by

$$F_X(x) = \begin{cases} r\dfrac{F_1(x)}{F_1(\theta)}, & 0 < x \leq \theta, \\ r + (1-r)\dfrac{F_2(x) - F_2(\theta)}{1 - F_2(\theta)}, & \theta < x < \infty. \end{cases}$$

We assume that $E(X^k) < \infty$ for some $k > 0$.
The cdf of the kth uncomplete moment is given by,

$$F_X^{(k)}(x) = \int_0^x \frac{t^k f_X(x)\, dx}{E(X^k)}, \quad 0 < x < \infty,$$

and the corresponding pdf,

$$f_X^{(k)}(x) = \frac{x^k f_X(x)}{E(X^k)}, \quad 0 < x < \infty.$$

The non-central moments of order k of the composite models are given in the following result (see Sarabia and Calderín-Ojeda, 2018):

Theorem 1.4 *The kth moment of the composite distribution with pdf defined in (1.24) is given by,*

$$E(X^k) = rE(X_1^k)\frac{F_1^{(k)}(\theta)}{F_1(\theta)} + (1-r)E(X_2^k)\frac{1 - F_2^{(k)}(\theta)}{1 - F_2(\theta)},$$

where $F_i^{(k)}(\cdot)$ denotes the kth uncomplete moment distribution of the random variable X_i, $i = 1, 2$.

Proof: The kth raw moment can be evaluated as,

$$E(X^k) = \int_0^\theta r\frac{x^k f_1(x)}{F_1(\theta)}\,dx + \int_\theta^\infty (1-r)\frac{x^k f_2(x)}{1 - F_2(\theta)}\,dx,$$

and using expression (1.24) we obtain directly (1.24). ∎

Example 1.8 *Let*

$$f_1(x) = \frac{1}{\sqrt{2\pi}\,x\,\sigma}\exp\left(-\frac{1}{2}\left(\frac{\ln x - \mu}{\sigma}\right)^2\right), \quad x > 0$$

be the pdf of a two-parameter lognormal distribution and

$$f_2(x) = \frac{\alpha\,\theta^\alpha}{x^{\alpha+1}}, \quad x > \theta,$$

be the pdf of a two-parameter Pareto distribution. Derive the composite lognormal-Pareto model.

Solution: The density function of this composite model is given by

$$f(x) = \begin{cases} r\dfrac{f_1(x)}{\Phi(\alpha\,\sigma)}, & 0 < x \le \theta, \\ (1-r)\,f_2(x), & \theta < x < \infty, \end{cases} \quad (1.24)$$

with $0 \le r \le 1$ and $\Phi(\cdot)$ denotes the cdf of the standard normal distribution. By allowing for continuity and differentiability at θ, we have that

$$r = \frac{\sqrt{2\pi}\,\alpha\,\sigma\,\Phi(\alpha\,\sigma)\exp\left(\tfrac{1}{2}(\alpha\sigma)^2\right)}{\sqrt{2\pi}\,\alpha\,\sigma\,\Phi(\alpha\,\sigma)\exp\left(\tfrac{1}{2}(\alpha\sigma)^2\right) + 1},$$

with

$$\alpha\sigma = \frac{\ln\theta - \mu}{\sigma},$$

where θ is the threshold, α is the tail index α and σ is an small loss parameter. □

1.4.3 General composite models

Previous results can be extended to the case of general composite models. Klugman et al. (2008) have considered the general form of the composite distributions in terms of the pdf function

$$f(x; \underline{r}, \underline{c}) = \begin{cases} r_1 f_1^*(x), & c_0 < x < c_1, \\ r_2 f_2^*(x), & c_1 < x < c_2, \\ \vdots & \vdots \\ r_k f_k^*(x), & c_{k-1} < x < c_k, \end{cases}$$

where,

$$f_i^*(x) = \frac{f_i(x)}{F_i(c_i) - F_i(c_{i-1})},$$

$i = 1, 2, \ldots, k$ with $\underline{r} = (r_1, \ldots, r_k)$ and $\underline{c} = (c_0, \ldots, c_k)$. These kind of distributions are also called k-component spliced distributions.

Many of the previous formulas can be extended to this general case. For example, the kth raw moment can be expressed as,

$$E(X^k) = \sum_{j=1}^{k} r_j E(X_j^k) \frac{F_j^{(k)}(c_j) - F_j^{(k)}(c_{j-1})}{F_j(c_j) - F_j(c_{j-1})}.$$

1.5 Multivariate distributions

In this topic, we will study the random variables multi-dimensional. The purpose of these is to measure two or more characteristics, such as the case of two or more risks related to the distribution of the number of claims in two years or to simultaneously study the expenditure made by a tourist at the origin and at the destination.

We will focus mainly on the study of two-dimensional random variables, although the resulting properties are easily extended to the case of more than two dimensions.

1.5.1 Bivariate Poisson distribution

The bivariate Poisson distribution plays a crucial role to analyse a diversity of phenomena in social sciences. For example, in actuarial statistics the insurer could be interested in modelling the joint number of claims in two different coverages in the insurance portfolio. In this section, we adopt the trivariate reduction method to construct the distribution (see, for instance, Johnson et al., 1997).

The probability mass function of the bivariate Poisson distribution is defined in the following way:

Definition 1.7 *Let* $X_i \sim Poisson(\theta_i)$, $i = 0, 1, 2$. *Let us consider the random variables*

$$X = X_1 + X_0,$$
$$Y = X_2 + X_0.$$

Then the random vector (X, Y) follows a bivariate Poisson distribution and it is denoted by
$$(X, Y) \sim \mathcal{BP}(\theta_1, \theta_2, \theta_0),$$
if its joint probability mass function is given by

$$\Pr(X = x, Y = y) = e^{-(\theta_1 + \theta_2 + \theta_0)} \frac{\theta_1^x}{x!} \frac{\theta_2^y}{y!} \sum_{i=0}^{\min(x,y)} \binom{x}{i}\binom{y}{i} i! \left(\frac{\theta_0}{\theta_1 \theta_2}\right)^i.$$

The marginal distributions of X and Y are given by
$$\begin{aligned} E(X) &= \theta_1 + \theta_0, \\ E(Y) &= \theta_2 + \theta_0, \end{aligned}$$
respectively.

Furthermore, $cov(X, Y) = \theta_0$, therefore θ_0 is a measure of dependence (positive correlation) between the two random variables. If $\theta_0 = 0$ then the two variables are independent and the bivariate Poisson distribution reduces to the product of two independent Poisson distributions.

The joint pgf and mgf of the random vector (X, Y) can be obtained by convolution. The pgf of the bivariate Poisson distribution is given by

$$G_{XY}(s, t) = e^{-\theta_1(s-1)} e^{-\theta_2(t-1)} e^{-\theta_0(ts-1)}, \text{ with } t, s \in \mathbb{R}.$$

On the other hand, the mgf of this bivariate distribution is

$$M_{XY}(s, t) = e^{\theta_1(e^s - 1)} e^{\theta_2(e^t - 1)} e^{\theta_0(e^t e^s - 1)}, \text{ with } t, s \in \mathbb{R}.$$

Finally the cumulant generating function of the bivariate Poisson is obtained by taking the log of (1.25):

$$K_{XY}(s, t) = \theta_1(e^s - 1) + \theta_2(e^t - 1) + \theta_0(e^t e^s - 1), \text{ with } t, s \in \mathbb{R}.$$

1.5.2 Bivariate Poisson distribution. An alternative parametrization

The bivariate Poisson distribution obtained via trivariate reduction is limited since it only accounts for positive correlation. In this section, we consider the bivariate Poisson proposed by Lakshminarayana et al. (1999) which allows for negative, zero or positive correlation. These authors introduced a mechanism to derive the bivariate Poisson distribution as a product of their marginal probability mass functions with a multiplicative factor or dependency parameter. This methodology can be easily extended to other discrete probability distributions.

1.5. MULTIVARIATE DISTRIBUTIONS

The probability mass function of the bivariate Poisson distribution is defined as follows:

Definition 1.8 *A random vector (X,Y) follows a bivariate Poisson distribution and it is represented as (X,Y) if its probability mass function is give as follows*

$$\begin{aligned} \Pr(X=x, Y=y) &= \Pr(X=x)\Pr(Y=y) \\ &\times \left[1 + \omega\left[\left(e^{-x} - e^{-c\theta_1}\right)\left(e^{-y} - e^{-c\theta_2}\right)\right]\right], \end{aligned}$$

where $c = 1 - e^{-1}$ and ω is the multiplicative factor parameter.

The marginal distributions of X and Y are Poisson with mean θ_1 and θ_2 respectively. The covariance between X and Y is given by

$$cov(X,Y) = \omega\sqrt{\theta_1\theta_2}c^2 e^{-\omega(\theta_1+\theta_2)}$$

and the correlation coefficient is

$$\rho = \frac{cov(X,Y)}{\sigma_x\sigma_y} = \omega\theta_1\theta_2 c^2 e^{-\omega(\theta_1+\theta_2)}.$$

Thus, the correlation coefficient can be positive, zero, or negative depending on the value of ω, the multiplicative factor parameter. Therefore, the correlation between the two random variables can be positive, zero, or negative.

The range of the dispersion parameter is

$$|\omega| \leq \frac{1}{(1-e^{-c\theta_1})(1-e^{-c\theta_2})},$$

therefore the correlation coefficient verifies the following inequality:

$$|\rho| \leq \frac{\theta_1\theta_2 c^2 e^{-\omega(\theta_1+\theta_2)}}{(1-e^{-c\theta_1})(1-e^{-c\theta_2})}.$$

The mgf of the random vector (X,Y) is

$$\begin{aligned} M_{X,Y}(t_1,t_2) &= e^{\theta_1(e^{t_1}-1)+\theta_2(e^{t_2}-1)} + \omega\left[e^{\theta_1(e^{t_1-1}-1)} - e^{-c\theta_1}e^{\theta_1(e^{t_1}-1)}\right] \\ &\times \left[e^{\theta_2(e^{t_2-1}-1)} - e^{-c\theta_2}e^{\theta_2(e^{t_2}-1)}\right]. \end{aligned}$$

This methodology can be easily implemented to derive the bivariate negative binomial distribution (see Famoye, 2010).

1.6 Multivariate continuous distributions

Two continuous multivariate distributions are illustrated in this chapter: The multivariate normal distribution and the multivariate (bivariate case) exponential distribution.

1.6.1 The multivariate normal distribution

The multivariate normal or Gaussian distribution, or joint normal distribution which is a generalization of the normal distribution to higher dimensions. This model is used to describe any set of correlated real-valued random variables, each of them clustered around its mean. One definition is that a random vector is said to be p-variate normally distributed if every linear combination of its p components has a univariate normal distribution. Properties of this multivariate family are examined.

The probability distribution of a pair of random variables is defined by continuous bivariate joint density. The most well-known continuous bivariate distribution is the bivariate normal distribution, which is made up of two independent random variables. The two variables in a bivariate normal are both are normally distributed, and they have a normal distribution when both are added together. The pictorial representation of this distribution is a three-dimensional bell curve. Many univariate tests and confidence intervals are based on the univariate normal distribution. Similarly, the majority of multivariate procedures have the multivariate normal distribution as their supporting structure.

The multivariate normal distribution of a p-dimensional random vector $\mathbf{X} = (X_1, \ldots, X_p)^\top$ can be written as follows $\mathbf{X} \sim N_p(\underline{\mu}, \mathbf{\Sigma})$, here $\underline{\mu}$ is a p-dimensional mean vector $\underline{\mu} = E(\mathbf{X}) = (E(X_1), \ldots, E(X_p))^\top$ and $\mathbf{\Sigma}$ is a $p \times p$ matrix, called the variance-covariance matrix, with entries $\Sigma_{ij} = E((X_i - \mu_i)(X_j - \mu_j)) = cov(X_i, X_j)$.

Definition 1.9 *The pdf of* $\mathbf{X}$ *in terms of* $\underline{\mu}$ *and* $\mathbf{\Sigma}$ *is given by*

$$f(x_1, \ldots, x_p) = \frac{1}{(2\pi)^{p/2}|\mathbf{\Sigma}|^{1/2}} \exp\left\{-\frac{1}{2}(\mathbf{x} - \underline{\mu})^\top \mathbf{\Sigma}^{-1}(\mathbf{x} - \underline{\mu})\right\},$$

where $|\mathbf{\Sigma}|$ *is the determinant of the variance-covariance matrix and* $\mathbf{x} = (x_1, \ldots, x_p)^\top \in \mathbb{R}^p$. *It only exists when* $\mathbf{\Sigma}$ *is positive-definite. where* $|\mathbf{\Sigma}|$ *is the determinant of the variance-covariance matrix. It only exists when* $\mathbf{\Sigma}$ *is positive-definite.*

Properties:

(i) Normality of linear combinations of the variables in $\mathbf{X}$. If the random vector $\mathbf{X} = (X_1, \ldots, X_p)^\top$ has a multivariate normal distribution, then the following results are satisfied:

Result 1.2 *Every linear combination* $Y = a_1 X_1 + \cdots + a_p X_p$ *of its component is normally distributed. That is,*

$$Y = a^\top \mathbf{X} = a_1 X_1 + \cdots + a_p X_p \sim N(a^\top \underline{\mu}, a^\top \mathbf{\Sigma} a),$$

1.6. MULTIVARIATE CONTINUOUS DISTRIBUTIONS

where $a = (a_1, \ldots, a_p)^\top$ is a constant vector with $a^\top \in \mathbb{R}^p$, $E(Y) = a^\top \underline{\mu}$, $\text{var}(Y) = a^\top \Sigma a$.

Result 1.3 *If A is a constant $q \times p$ matrix of rank q, where $q \leq p$, the q linear combinations in $A\mathbf{X}$ have a multivariate normal distribution: If $\mathbf{X}$ is $N_p(\underline{\mu}, \Sigma)$, then $A\mathbf{X}$ is $N_q(A\underline{\mu}, A\Sigma A^\top)$.*

(*ii*) Chi-square distribution:

Result 1.4 *A chi-square random variable with p degrees of freedom is defined as the sum of squares of p independent standard normal random variables. If $\mathbf{X}$ is $N_p(\underline{\mu}, \Sigma)$ and Σ is positive-definite, then $(\mathbf{X} - \underline{\mu})^\top \Sigma^{-1} (\mathbf{X} - \underline{\mu}) \sim \chi^2(p)$.*

(*iii*) Normality of marginal distributions:

Result 1.5 *If $\mathbf{X}$ is $N_p(\underline{\mu}, \Sigma)$, then X_j is $N(\mu_j, \sigma_{jj})$, $j = 1, 2, \ldots, p$.*
The converse of this is not true. If the density of each X_j is normal, it does not necessarily follow that $\mathbf{X}$ is multivariate normal.

(*iv*) Independence:

Result 1.6 *Two individual variables X_j and X_k are independent if $\sigma_{jk} = 0$. Note that this is not true for many non-normal random variables.*

(*v*) Conditional distribution: Let the observation vector be partitioned into two sub vectors denoted by $\mathbf{Y}$ and $\mathbf{Z}$, where $\mathbf{Y}$ is $p \times 1$ and $\mathbf{Z}$ is $q \times 1$. Then,

$$E\begin{pmatrix}\mathbf{Y}\\ \mathbf{Z}\end{pmatrix} = \begin{pmatrix}\underline{\mu}_Y\\ \underline{\mu}_Z\end{pmatrix}, \quad \text{cov}\begin{pmatrix}\mathbf{Y}\\ \mathbf{Z}\end{pmatrix} = \begin{pmatrix}\Sigma_{YY} & \Sigma_{YZ}\\ \Sigma_{ZY} & \Sigma_{ZZ}\end{pmatrix}.$$

It is assumed that

$$\begin{pmatrix}\mathbf{Y}\\ \mathbf{Z}\end{pmatrix} \sim N_{p+q}\left(\begin{pmatrix}\underline{\mu}_Y\\ \underline{\mu}_Z\end{pmatrix}, \begin{pmatrix}\Sigma_{YY} & \Sigma_{YZ}\\ \Sigma_{ZY} & \Sigma_{ZZ}\end{pmatrix}\right).$$

Result 1.7 *If $\mathbf{Y}$ and $\mathbf{Z}$ are not independent, then $\Sigma_{\mathbf{ZY}} \neq \mathbf{O}$, and the conditional distribution of $\mathbf{Y}$ given $\mathbf{Z}$, and $f(\mathbf{Y}|\mathbf{Z})$, is multivariate normal with*

$$\begin{aligned} E(\mathbf{Y}|\mathbf{Z}) &= \underline{\mu}_Y + \Sigma_{YZ}\Sigma_{ZZ}^{-1}(\mathbf{Z} - \underline{\mu}_Z),\\ \text{cov}(\mathbf{Y}|\mathbf{Z}) &= \Sigma_{YY} - \Sigma_{YZ}\Sigma_{ZZ}^{-1}\Sigma_{ZY}. \end{aligned}$$

(vi) Distribution of the sum of two subvectors:

Result 1.8 *If $\mathbf{U}$ and $\mathbf{V}$ are the same size (both $p \times 1$) and independent, then*

$$\mathbf{U} + \mathbf{V} \sim N_p(\underline{\mu}_U + \underline{\mu}_V, \mathbf{\Sigma_{UU}} + \mathbf{\Sigma_{VV}}),$$
$$\mathbf{U} - \mathbf{V} \sim N_p(\underline{\mu}_U - \underline{\mu}_V, \mathbf{\Sigma_{UU}} + \mathbf{\Sigma_{VV}}).$$

If $p = 2$, the bivariate normal distribution is obtained. The bivariate normal distribution can be defined as the pdf of two variables X and Y that are linear functions of the same independent normal random variables:

$$f(x, y) = \frac{1}{2\pi\sigma_1\sigma_2\sqrt{1-\rho^2}} \exp\left\{-\frac{1}{2(1-\rho^2)} Q(x, y)\right\}, \qquad (1.25)$$

where

$$Q(x, y) = \left(\frac{x - \mu_x}{\sigma_x}\right)^2 - 2\rho \left(\frac{x - \mu_x}{\sigma_x}\right)\left(\frac{y - \mu_y}{\sigma_y}\right) + \left(\frac{y - \mu_y}{\sigma_y}\right)^2,$$

being ρ the correlation coefficient.

In Figure 1.13, we plot the pdf (left panel) and contour plot (right panel) of the bivariate normal distribution (1.25) with means $\mu_1 = 1$, $\mu_2 = 1.5$, standard deviations $\sigma_1 = 0.5$, $\sigma_2 = 0.4$ and correlation coefficient $\rho = 0.7$.

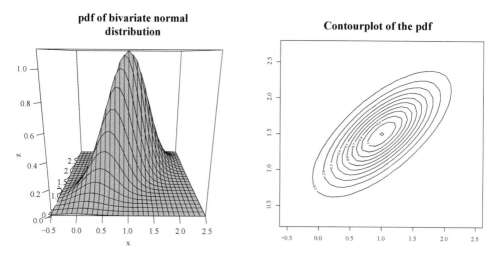

Figure 1.13: Probability density function (left panel) and contour plot (right panel) of the bivariate normal distribution (1.25) with parameter values $\mu_1 = 1$, $\mu_2 = 1.5$, $\sigma_1 = 0.5$, $\sigma_2 = 0.4$ and $\rho = 0.7$.

1.6.2 Bivariate exponential distribution

The exponential distribution plays a key role to explain a high number of events that occur in nature. For example, it is used as a baseline model for life testing. It has also used as the starting point for the theory of extreme values. It may therefore be of interest to derive bivariate distributions where the marginal distributions are exponential (see Gumbel, 1960). The bivariate exponential distribution is defined in terms of its cdf as follows:

Let $F_X(x)$ and $F_Y(y)$, $f_X(x)$ and $f_Y(Y)$ be the probability and density functions of continuous random variables X and Y.

Definition 1.10 *The cdf of the bivariate exponential distribution is given by*

$$F(x,y) = 1 - e^{-x} - e^{-y} + e^{-x-y-\gamma xy}, \text{ with } x, y \geq 0, \text{ and } 0 \leq \gamma \leq 1.$$

At the boundary, it is satisfied that $F(x,0) = F(0,y) = F(0,0) = 0$ and $F(\infty, \infty) = 1$. When $\gamma = 0$, the random variables are independent and it is verified that $F(x,y) = F_X(x)F_Y(y)$.

If the second cross partial derivative $\partial^2 F/\partial x \partial y$ exists everywhere, the bivariate distribution has a density $f(x,y)$ equal to this derivative and the condition that the probability of every rectangle is non-negative.

The pdf of the bivariate exponential distribution is given by

$$f(x,y) = e^{-x(1+\gamma y)-y}[(1+\gamma x)(1+\gamma y) - \gamma].$$

It is verified that $f(\infty, y) = f(x, \infty) = 0$ and $f(0,0) = 1 - \delta$.

Properties:

- It is verified that $F(x,y) \leq F_X(x)$ and $F(x,y) \leq F_Y(y)$.

  Proof: As $F(x,y)$ can be written as

  $$F(x,y) = 1 - e^{-x} - e^{-y}(1 - e^{-x-\gamma xy}),$$

  and assuming that $0 \leq \gamma \leq 1$, the result follows immediately. ∎

- The conditional density are given by the following expressions

  $$\begin{aligned} f(x|y) &= e^{-x(1+\gamma y)}[(1+\gamma x)(1+\gamma y) - \gamma], \\ f(y|x) &= e^{-y(1+\gamma x)}[(1+\gamma x)(1+\gamma y) - \gamma]. \end{aligned}$$

- The covariance of the bivariate is given by

  $$\text{cov}(X,Y) = -\frac{e^{1/\gamma}}{\gamma} Ei(-1/\gamma) - 1,$$

  where $Ei(\cdot)$ is the integrand logarithm.

- The coefficient of correlation, $\rho = \text{cov}(X,Y)$, is zero for $\gamma = 0$ and it decreases for increasing values of γ up to $\rho = -0.40365$, for $\gamma = 1$.

1.7 Criteria for model validation

In this section, we will provide some statistical tools for model validation and therefore for model selection.

1.7.1 Hypothesis testing

Hypothesis tests are an essential tool for selecting and validating the different models that can fit the available data well.

Chi-square test

In this context, the most widely used statistical test for validation of a model is the χ^2 test of goodness of fit. This test aims to decide if the empirical distribution follows a given distribution from the observed data. It allows us to test if the observed data corresponds to a model where the data is chosen from the observed values according to some specified set of probabilities obtained from the theoretical distribution. We can probabilistically quantify the discrepancy between what you observe and what you expect to observe.

Let us suppose that n observations in a random sample from a population are classified into k mutually exclusive classes with corresponding observed numbers denoted by o_i for $i = 1, 2 \ldots, k$, and under the null hypothesis we assume that the probability that an observation falls into the ith class is denoted by p_i, then we have that the expected numbers $e_i = n\, p_i$, where $\sum_{i=1}^{k} p_i = 1$ and $n = \sum_{i=1}^{k} o_i = \sum_{i=1}^{k} e_i$, then the statistics

$$\chi^2 = \sum_{i=1}^{k} \frac{(o_i - e_i)^2}{e_i}$$

when n is large, it is asymptotically distributed as a chi-square distribution with $n - 1$ degrees of freedom. In general, if we need to estimate p parameters in the theoretical distribution it follows as a chi-square distribution with $n - p - 1$ degrees of freedom.

Tests based on the empirical distribution function

It is also natural to express the fit of the theoretical model to the data in terms of distribution functions. In particular, it is suggested to use the following three empirical distribution function goodness-of-fit measures to quantify the distance between the empirical distribution function constructed from the data and the cdf of the fitted models, namely the Kolmogorov-Smirnov (KS), Cramer-von Mises (CVM) and Anderson-Darling (AD) test statistic (see Rizzo, 2009). The definition of the test statistics are given as follows: Denote the cdf function of the fitted model by $\hat{F}(\cdot)$, the original data by $x_1, \ldots, x_n$ and the ordered data in increasing magnitude by $x_{(1)}, \ldots, x_{(n)}$, then we have:

1.7. CRITERIA FOR MODEL VALIDATION

(i) Kolmogorov-Smirnov (KS) test statistics: $D = \max(D^+, D^-)$, where

$$D^+ = \max_{1 \leq j \leq N} \left\{ \frac{j}{N} - \hat{F}(x_{(j)}) \right\}, D^- = \max_{1 \leq j \leq N} \left\{ \hat{F}(x_{(j)}) - \frac{j-1}{N} \right\}$$

(ii) Cramer-von Mises (CVM) test statistics:

$$W^2 = \sum_{j=1}^{N} \left[\hat{F}(x_{(j)}) - \frac{2j-1}{2N} \right]^2 + \frac{1}{12N}$$

(iii) Anderson-Darling (AD) test statistics:

$$A^2 = -N - \frac{1}{N} \sum_{j=1}^{N} [(2j-1)\log(\hat{F}(x_{(j)})) + (2n+1-2j)\log(1-\hat{F}(x_{(j)}))]$$

For all the empirical distribution function goodness-of-fit measures above, the smaller values indicate a better fit of the model to the data.

Vuong's test

This test proposed by Vuong (see Vuong, 1989) is also considered as a tool for model diagnostic. This statistic makes probabilistic statements about two models that can be nested, non-nested or overlapping. The statistic tests the null hypothesis that the two models are equally close to the actual data generating process, against the alternative that one model is closer. The test statistic is

$$T = \frac{1}{\omega \sqrt{n}} \left(\ell_f(\hat{\theta}_1) - \ell_g(\hat{\theta}_2) - \log n \left(\frac{p}{2} - \frac{q}{2} \right) \right),$$

where

$$\omega^2 = \frac{1}{n} \sum_{i=1}^{n} \left[\log \left(\frac{f(\hat{\theta}_1)}{g(\hat{\theta}_2)} \right) \right]^2 - \left[\frac{1}{n} \sum_{i=1}^{n} \log \left(\frac{f(\hat{\theta}_1)}{g(\hat{\theta}_2)} \right) \right]^2 \quad (1.26)$$

is the estimated variance calculated in the usual manner and $f(\cdot)$ and $g(\cdot)$ represent the pdf of two different non-nested models, $\hat{\theta}_1$ and $\hat{\theta}_2$ are the maximum likelihood estimates of the parameters θ_1 and θ_2 and p and q are the number of estimated coefficients in the model with pdf f and g respectively. Note that the Vuong's statistic is sensitive to the number of estimated parameters in each model and therefore the test must be corrected for dimensionality. Under the null hypotheses, $H_0 : E(\ell_f(\hat{\theta}_1) - \ell_g(\hat{\theta}_2)) = 0$, T is asymptotically normally distributed. It is generally accepted that the rejection region for this test in favor of the alternative hypothesis occurs, at the 5% significance level, when $T > 1.96$.

1.7.2 Other measures of model selection

Model assessment can be also examined from theoretical plausibility. In this case, the model selection is justified employing Kullback-Leibler divergence, suggesting an information-criterion based approach. The following five information criteria are used:

(i) Maximum of the negative log-likelihood (NLL): Calculated by taking the negative of maximum value of the log-likelihood function evaluated at the maximum likelihood estimates.

(ii) Akaike information criterion (AIC): Calculated by twice the NLL, evaluated at the maximum likelihood estimates, plus twice the number of estimated parameters.

(iii) Bayesian information criterion (BIC): Obtained as twice the NLL, evaluated at the maximum likelihood estimates, plus $k\ln(n)$, where k is the number of estimated parameters and n is the sample size.

(iv) Consistent Akaike Information Criteria (CAIC): A corrected version of the AIC, proposed by Bozdogan (1987) to overcome the tendency of the AIC overestimating the complexity of the underlying model as it lacks the asymptotic property of consistency. To calculate the CAIC, a correction factor based on the sample size is used to compensate for the overestimating nature of AIC. The CAIC is defined as twice the NLL plus $k(1+\ln(n))$, again k is the number of free parameters and n refers to the sample size.

(v) Hanann-Quinn information criterion (HQIC): This measure is computed by twice the NLL plus $2(d+1)\log(\log n)$ where d is the number of parameters in the model and n is the sample size (see Hannan and Quinn, 1979).

Note that for all the information criterion above, smaller values indicate a better fit of the model to the data.

1.7.3 Graphical methods of model selection

It is usual to perform a graphical analysis of the data to validate the fit to data of several models. Among the best known, we will describe the Q-Q plots (quantile-quantile plots). Sometimes it is useful to plot these graphs in log scale (log-log plots) to show a linear relationship between the observed data and their corresponding ranks. This is is useful in different areas of economics such as urban economics or income and wealth distributions.

Q-Q plots

Let us suppose that we have a sample $x_1, \ldots, x_n$ from a population with cdf that depends on a certain parameter λ given by $F(x; \lambda)$ and let us also consider that the

1.7. CRITERIA FOR MODEL VALIDATION

ordered sample $x_{(1)} \leq x_{(2)} \leq \cdots \leq x_{(n)}$ and the points $p_{(i)} = \frac{i}{n+1}, i = 1, \ldots, n$. Let $F(x; \hat{\lambda})$ be the cdf evaluated at an estimate obtained from the sample $\mathbf{x} = (x_1, \ldots, x_n)$, such that $F^{-1}(p_{(n)}; \hat{\lambda})$ is the estimated quantile, i.e., it corresponds to the order statistic i, $x_{(i)}$.

Definition 1.11 *We define the Q-Q plot as the resulting graph of plotting the pairs* $(x_{(i)}, F^{-1}(p_{(i)}; \hat{\lambda})), i = 1, \cdots, n$.

The Q-Q plot shows the observed and estimated quantiles, then if the model fits the data correctly, the resulting scatter plot must be located on the diagonal of the rectangle $[F(p_{(1)}; \hat{\lambda}), F(p_{(n)}; \hat{\lambda})] \times [x_{(1)}, x_{(n)}]$, which is the graphical representation of the Q-Q plot.

These graphs can be also drawn in log–log scale where the complementary of the cdf is plotted against the observed ordered data.

Example 1.9 *Simulate 1000 random variates from a lognormal distribution with location parameter $\mu = 0.5$ and $\sigma = 1$. Estimate the parameters by the method of maximum likelihood and sketch the Q-Q plot for this model.*

Solution: The following code developed in Mathematica v.12 provides the maximum likelihood estimates for the lognormal distribution with parameters $\mu = 0.5$ and $\sigma = 1$:

```
Clear[x, logl, LH, f, mu, sigma];
DATA = RandomVariate[LogNormalDistribution[0.5, 1], 1000];
f[x_, mu_, sigma_] :=
  PDF[LogNormalDistribution[mu, sigma], x];
LH[mu_,sigma_] :=
  Sum[Log[f[DATA[[i]], mu, sigma]], {i, 1, 1000}];
veros = LH[mu, sigma];
logl = FindMaximum[veros, {{mu, 0.5}, {sigma, 1.00}},
  MaxIterations -> 1000]
mle = logl[[2]];
```

The estimates of the parameters are $\hat{\mu} = 0.4909$ and $\hat{\sigma} = 1.0256$. Also, the code to compute the quantiles is given below:

```
Clear[p];
datos1 = Table[Quantile[DATA, p], {p, 0.001, 0.999, 0.001}];
mu = mle[[1, 2]]; sigma = mle[[2, 2]];
quant = InverseCDF[LogNormalDistribution[mu, sigma], p];
datos2 = Table[quant, {p, 0.001, 0.999, 0.001}];
```

The Q-Q plot for this model is given in Figure 1.14. The sample quantiles are 1000 random variates generated from lognormal with $\mu = 0.5$ and $\sigma = 1$ whereas the

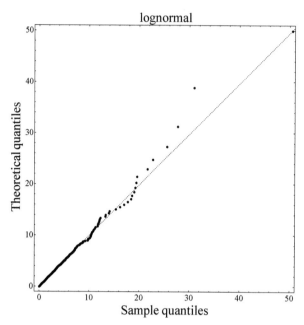

Figure 1.14: Q-Q plot. Sample quantiles are obtained from 1000 random variates from lognormal with $\mu = 0.5$ and $\sigma = 1$. Theoretical quantiles are obtained from lognormal with $\hat{\mu} = 0.4909$ and $\hat{\sigma} = 1.0256$.

theoretical quantiles are obtained from lognormal with $\hat{\mu} = 0.4909$ and $\hat{\sigma} = 1.0256$. It is observable that the theoretical quantiles adhere perfectly well to the observed quantile in most of the graph and the theoretical quantiles are more significant than the empirical quantile in the top part of the graph.

□

Exercises

1. The geometric distribution also gives the probability that the first occurrence of success requires n independent trials, each with success probability p. The probability that the nth trial is the first success is

$$\Pr(N = n) = (1-p)^{n-1}p, \quad n = 1, 2, \ldots$$

   a) Show that the mgf of N is

$$M_N(t) = \frac{pe^t}{1 - (1-p)e^t}, \text{ for } t < \log 1/(1-p).$$

1.7. CRITERIA FOR MODEL VALIDATION

b) Show that the mean and variance are given by

$$E(X) = \frac{1}{p},$$
$$var(X) = \frac{1-p}{p^2},$$

respectively.

2. The Weibull distribution also has pdf given by

$$f(y|\tau, m) = \begin{cases} \dfrac{1}{\tau} m\, y^{m-1} e^{-\frac{y^m}{\tau}}, & y > 0, \\ 0, & \text{Otherwise.} \end{cases}$$

where α and m are positive constants.

a) Show that this density is obtained from (1.13) by taking $\alpha = m$ and $\tau = \sigma^m$;

b) Proof that the random variable $U = Y^m$ is exponentially distributed with mean τ.

Find the density function of $U = Y^m$.

3. Let $X|\theta$ have an exponential distribution with mean θ. Let Θ have a gamma distribution with shape parameter α and rate parameter λ. Prove that the unconditional distribution of X is Pareto type II with shape parameter α and scale parameter λ.

4. Let us suppose that the number of claims X, follows a Poisson distribution with mean $z > 0$. Let us also suppose that this parameter follows a reciprocal inverse Gaussian distribution (see Gómez-Déniz and Calderín-Ojeda, 2018) with density function given by

$$g(z) = \frac{\gamma}{\sqrt{2\pi z}} \exp\left[-\frac{(\gamma z - \lambda)^2}{2z}\right], \quad z > 0, \; \gamma > 0, \; \lambda > 0.$$

Show that the unconditional distribution of X is provided by

$$\Pr(X = x) = \frac{\gamma e^{\gamma \lambda}}{x!} \sqrt{\frac{2}{\pi}} \left(\frac{\lambda}{\sqrt{2+\gamma^2}}\right)^{1/2+x} K_{1/2+x}(\lambda \phi),$$
$$x = 0, 1, \ldots,$$

where $K(\cdot)$ is the modified Bessel function of the third kind.

5. The mean and variance of (1.27) are μ and $\mu + \frac{1}{\gamma^2}\left(\mu + \frac{1}{\gamma^2}\right)$. Show that the probability mass function given by (1.27) is overdispersed, i.e., the variance is larger than the mean.

6. The returns of two stocks can be modelled via a bivariate normal distribution $(X_1, X_2)^\top$ with means $\mu_1 = 0.5$, $\mu_2 = 0.2$, standard deviations $\sigma_1 = 0.15$, $\sigma_2 = 0.25$ and covariance $\sigma_{12} = 0.03$. Calculate:

   a) Show that $Y = 0.64X_1 + 0.36X_2$ follows $N(0.392, 0.0311)$;

   b) Check that $\Pr(0.64X_1 + 0.36X_2 > 0.53)$ is 0.2171;

   c) Obtain the same result by simulation.

7. The daily price of two shares $(X_1, X_2)^\top$ follows a bivariate normal distribution with means 4.1 and 5.2, variances 2 and 2.5 and correlation coefficient 0.58:

   a) Check that $\Pr\left(X_1 > \dfrac{X_1 + X_2}{2}\right)$ is 0.2128.

   b) Show that the simulated values of the median and the first and third quartiles of the distribution of the maximum daily price of X_1 and X_2 are 4.36, 5.35 and 6.35 respectively.

8. Show that the mode of the boundary density function $f(x|0)$ associated with the bivariate exponential distribution is $x_{Mo} = 2 - 1/\gamma$.

9. Show that the marginal PGF for the bivariate Poisson distribution, satisfies the following expression $G_X(s) = G_{XY}(s, 1)$.

10. Show that the moment estimators of the bivariate Poisson distributions given a sample of size n, $\{(x_1, y_1), \ldots, (x_n, y_n)\}$ are

$$\hat{\lambda}_1 = \bar{x}, \quad \hat{\lambda}_2 = \bar{y}, \quad \hat{\omega} = \frac{\sum_{i=1}^n (x_i - \bar{x})(y_i - \bar{y})}{n\lambda_1 \lambda_2 e^{-c(\lambda_1 + \lambda_2)}(e - e^{-1})^2}.$$

Chapter 2

Statistical Distributions in Insurance and Finance

2.1 Introduction

In this chapter, we study the application of the statistical distributions presented in the previous chapter to insurance and finance. In the second subsection we will discuss two different methodologies to model aggregate claims in a portfolio: the individual risk model and the collective risk model. The former approach deals with the claims on individual policies and summing all of them in the portfolio. On the other hand, the latter approach is derived from the portfolio as a whole entity. Here we provide formulae for the considered premiums in the collective risk model. Also, as for the random variable of the number of claims N, the Poisson or a negative binomial distribution is often selected, we discuss these cases in detail. In the following section, we will introduce two important families of counting distributions that will be useful to calculate aggregate claims distribution, they are the $(a, b, 0)$ class and the $(a, b, 1)$ class. Next, a recursive expression for the aggregate claims distribution is presented. When the claims number distribution belongs to the $(a, b, 0)$ class of distributions, Panjer's recursion formula allows deriving a recursive calculation of the amount of the aggregate claims when the individual's claims sizes are distributed on the non-negative integers. The advantage of this expression is that there is no need to compute convolutions in the calculation of the probability function of the aggregate claim amount. Later, we provide and examine several mathematical methods to calculate premiums. A premium is a payment that an insured makes for total or partial cover against risk. More formally, the premium can be defined as the amount that the policyholder pays for insurance coverage. Premium calculation principles can be considered as special cases of risk measures, for that reason risk measures are discussed. In general, risk measures are used for determining provisions and capital requirements to avoid insolvency. Since risks are modelled as non-negative random variables, measuring risk is equivalent to establishing a correspondence between the space of random variables and $\mathbb{R}^+$. In this section, we will study risk measures that measure the upper tails of distribution functions and introduce the concept of reinsurance. Reinsurance is the procedure by which insurers transfer part of their risk portfolios to other parties by some form of contract to reduce the probability of paying a significant obligation due to a claim. The party that diversifies its insurance portfolio is the ceding company or insurer. The party that accepts a portion of the potential obligation in exchange for a share of the insurance premium is known as the reinsurer. Broadly speaking, reinsurance is known as insurance for insurers. Reinsurance allows insurers to remain solvent by recovering some or all amounts paid to claimants. Reinsurance reduces the net liability on individual risks and catastrophe protection from significant or multiple losses by paying to the reinsurer a reinsurance premium. This process provides the ceding companies the capacity to increase their underwriting possibilities in terms of the number and size of risks. In the second part of this section, we discuss the different type of reinsurance contracts. In the last section of this chapter, a comparison of risks is considered. Ordering of risks is crucial in aspect in the actuarial profession. There are two reasons why a risk, that is, a non-negative random loss would be preferred to another: On the one hand, the other risk is larger, on the other hand, the other risk is riskier. This second motivation is related to the random variables with a thicker tail. The latter reason means that the probability of extreme values is larger, making a risk with equal mean less attractive because it is more spread, i.e., less predictable. Having thicker tails means having larger stop-loss premiums. From the fact that one risk is smaller than another, one may deduce that it is also preferable in the mean-variance order. In this ordering, the practitioner prefers the risk with the smaller mean, and the variance is used as a tie-breaker. However, this ordering is inadequate for actuarial purposes, since it leads to wrong decisions. Definitions of stochastic dominance provided and different associated results are examined. The relation of stochastic dominance and optimal reinsurance is also considered.

2.2 Individual and collective risk models

In this section, we consider two different approaches to examine the aggregate claims arising from a general insurance portfolio over a short period: the individual risk model and the collective risk model.

2.2.1 Individual risk model

The individual risk model considers the number of claims arising from a portfolio as the sum of the individual claim amounts from the individual policies in the portfolio. In the following, we specify this model.

Let us consider n independent policies $X_1, \ldots, X_n$ where n is a fixed quantity and represent the total number of policies in the portfolio. The total claims on the portfolio is given by S where $S = X_1 + \cdots + X_n$ is a fixed sum, not a random sum,

2.2. INDIVIDUAL AND COLLECTIVE RISK MODELS

since n is not a random variable. It is important to note that some of the X_i with $i = 1, \ldots, n$ can be zero.

Then, by independence we have

$$E(S) = \sum_{i=1}^{n} X_i \text{ and } var(S) = \sum_{i=1}^{n} var(X_i).$$

Let us now assume that $C_1, \ldots, C_z$ are the claims on policies on which there is a claim, we have that $S = C_1 + \cdots + C_z$ with $z < n$. Let us suppose that $C_1, \ldots, C_z$ are independent, identically distributed random variables, and that C_i with $i = 1, \ldots, z$ has mgf $M_C(t)$. Suppose also that z is a realisation of a random variable Z, and $Z, C_1, \ldots, C_z$ are mutually independent and that Z has mgf $M_Z(t)$. Then the random variable $S = C_1 + \cdots + C_z$ is a random sum with mgf given by

$$M_S(t) = E(e^{tS}) = E(E(e^{tS}|Z = z)),$$

and by the law of iterated expectations

$$\begin{aligned} E(E(e^{tS}|Z=z)) &= E(e^{t(C_1 + \cdots + C_z)}) \\ &= \prod_{i=1}^{z} E(e^{tC_i}) = M_C(t)^z. \end{aligned}$$

Therefore,

$$\begin{aligned} M_S(t) &= E(M_C(t)^Z) \\ &= E(e^{Z \log M_C(t)}) = M_Z(\log M_C(t)). \end{aligned}$$

Let us also assume that $X_i = I_i C_i$, where I_i is and indicator that takes the value 1, if there is a claim on policy i, and 0 otherwise. Then, the probability of claim on policy i is given by

$$\Pr(I_i = 1) = \Pr(\text{claim on policy } i) = p \text{ independent of } i.$$

Therefore, I_i is a Bernoulli random variable and Z has a binomial distribution with parameters n and p with mgf given by

$$\begin{aligned} M_Z(t) &= E(e^{Zt}) = \sum_{z=0}^{n} e^{zt} \binom{n}{z} p^z (1-p)^{n-z} \\ &= (1 - p + pe^t)^n. \end{aligned}$$

Consequently, we have that

$$\begin{aligned} M_S(t) &= E(e^{Zt}) = \sum_{z=0}^{n} e^{zt} \binom{n}{z} p^z (1-p)^{n-z} \\ &= (1 - p + pe^t)^n. \end{aligned}$$

Thus,

$$M_S(t) = (1 - p + pM_C(t))^n. \tag{2.1}$$

The cumulant generating function of S is computed by taking the log of (2.1). Its expression is given by

$$K_S(t) = \log E(e^{Zt}) = n\log(1 - p + pM_C(t)). \tag{2.2}$$

Now, assuming that the mean and variance of C_i are μ and σ^2, and differentiating the expression (2.2) and evaluating at 0, we obtain the mean of S,

$$E(S) = \left.\frac{dK_S(t)}{dt}\right|_{t=0} = \frac{npM'_C(0)}{1 - p + pM_C(0)} = np\mu,$$

since $M'_C(0) = \mu$ and $M_C(0) = 1$.

Now differentiating twice the expression (2.2) and evaluating at 0, we obtain the variance of S,

$$var(S) = \left.\frac{d^2K_S(0)}{dt^2}\right|_{t=0} = n(p\sigma^2 + p(1-p)\mu^2).$$

Example 2.1 *Give the expression for the mean and variance of the total claims amount S under the individual risk model when the claims size follow an exponential distribution with scale parameter λ.*

Solution: By using the expression (2.3), we have that the mean of S is given by

$$E(S) = np\lambda.$$

Similarly using (2.3), we have that

$$var(S) = n\lambda^2 p(2 - p).$$

□

2.2.2 Collective risk model

The collective risk model quantify the aggregate claims in an insurance portfolio by considering the portfolio as a whole entity. This is defined below:

Definition 2.1 *Let us consider a fixed period of time, e.g., one year. Let*

$$S = \sum_{i=1}^{N} X_i,$$

with $S = 0$ if $N = 0$, where

(i) X_i is the size of the ith claim,

2.2. INDIVIDUAL AND COLLECTIVE RISK MODELS

(ii) N is the number of claims, and

(iii) S is the aggregate claim amount.

This model is defined as the the collective risk model.

The model assumptions are:

i) $\{X_i\}_{i=1}^{\infty}$ is a sequence of non-negative and independent and identically distributed random variables with cdf $F(\cdot)$. We will assume $F(0) = 0$ and $\mathrm{E}(X_1^k) = m_k$.

ii) N is independent of $\{X_i\}_{i=1}^{\infty}$.

Theorem 2.1 *The cdf of S is given by*

$$G(x) = \Pr(S \leq x) = \sum_{n=0}^{\infty} F^{n*}(x) \Pr(N = n),$$

where $F^{n*}(\cdot)$ is the n-fold convolution of $F(\cdot)$, with $F^{1*}(x) = F(x)$ and $F^{0*}(x) = 1$ and $G(0) = \Pr(N = 0)$.

Proof: Let $G(x) = \Pr(S \leq x)$, then

$$\begin{aligned}
\Pr(S \leq x) &= \sum_{n=0}^{\infty} \Pr(S \leq x \text{ and } N = n) \\
&= \sum_{n=0}^{\infty} \Pr(S \leq x) | N = n) \Pr(N = n) \\
&= \sum_{n=0}^{\infty} F^{n*}(x) \Pr(N = n).
\end{aligned}$$

As we have assumed that $F(0) = 0$, we have $\Pr(S = 0) = \Pr(N = 0)$. ∎

Remark 2.1 *When $F(\cdot)$ is a continuous distribution with $F(0) = 0$, S has a mixed distribution with density function*

$$g(x) = \frac{d}{dx} G(x) = \sum_{n=1}^{\infty} f^{n*}(x) \Pr(N = n),$$

for $x > 0$, and with a mass of probability at 0.

Theorem 2.2 *The mean, variance and mgf of the random variable S are given by*

$$\begin{aligned}
E(S) &= E(N) m_1, \\
var(S) &= E(N)(m_2 - m_1^2) + var(N) m_1^2, \\
M_S(t) &= M_N(\log M_X(t)).
\end{aligned}$$

Proof: First, by using the law if iterated expectations we have:

$$E(S) = E(E(S|N))$$

and

$$E(S|N) = E\left(\sum_{i=1}^{n} X_i\right) = nm_1,$$

so that

$$\begin{aligned} E(S) &= E(N)m_1, & (2.3) \\ var(S) &= E(var(S|N)) + var(E(S|N)), & (2.4) \end{aligned}$$

and

$$var(S|N=n) = var\left(\sum_{i=1}^{n} X_i\right) = n\,var(X_1) = n(m_2 - m_1^2),$$

therefore

$$var(S) = E(N(m_2 - m_1^2)) + var(Nm_1) = E(N)(m_2 - m_1^2) + var(N)m_1^2.$$

Finally,

$$M_S(t) = E(e^{tS}) = E(E(e^{tS}|N)),$$

and

$$\begin{aligned} E(E(e^{tS}|N=n)) &= E(e^{t(X_1+\cdots+X_n)}) \\ &= \prod_{i=1}^{n} E(e^{tX_i}) = M_X(t)^n. \end{aligned}$$

Therefore,

$$\begin{aligned} M_S(t) &= E(M_X(t)^N) \\ &= E(e^{N \log M_X(t)}) = M_N(\log M_X(t)). \end{aligned}$$

∎

Corolario 2.1 *If $\{X_i\}_{i=1}^{\infty}$ is a sequence of non-negative and discrete independent and identically distributed random variables, the pgf of S is*

$$P_S(s) = P_N(P_X(s))$$

2.2. INDIVIDUAL AND COLLECTIVE RISK MODELS

2.2.3 Compound Poisson distribution

When $N \sim \mathcal{P}(\lambda)$ we say that S has a compound Poisson distribution with parameters λ and the claims sizes have common cdf given by $F(\cdot)$.

As $E(N) = var(N) = \lambda$, we have that $E(S) = \lambda m_1$ and $var(S) = \lambda m_2$. Also as

$$M_N(t) = \exp\{\lambda(e^t - 1)\}$$

then,

$$M_S(t) = \exp\{\lambda(M_X(t) - 1)\}.$$

Now, from the cumulant generating function, the central moments can be obtained, by using the expression,

$$\frac{d^n}{dt} K_S(t)_{|t=0} = E((S - E(S))^n),$$

where $K_S(t) = \log M_S(t) = \lambda(M_X(t) - 1)$.

For example, the third central moment is

$$E((S - E(S))^3) = \lambda m_3,$$

the compound Poisson distribution is positively skewed with coefficient of skewness $\frac{\lambda m_3}{\sqrt[3]{(\lambda m_2)^2}}$.

2.2.4 Compound negative binomial distribution

When $N \sim \mathcal{NB}(r, p)$ we say that S has a compound negative binomial distribution with parameters r and p and the claims sizes have common cdf given by $F(\cdot)$.

As $E(N) = \frac{rp}{1-p}$ and $var(N) = \frac{rp}{(1-p)^2}$, we have that $E(S) = \frac{rp}{1-p} m_1$ and $var(S) = \frac{rp}{1-p} \left(m_2 + m_1^2 \frac{p}{1-p} \right)$. Also as

$$M_N(t) = \frac{p^r}{(1 - (1-p)e^t)^r},$$

given that $0 < (1-p)e^t < 1$, or $t < -\log(1-p)$. Then,

$$M_S(t) = \frac{p^r}{(1 - (1-p)M_X(t))^r},$$

given that $0 < (1-p)M_X(t) < 1$, or $M_X(t) < 1/(1-p)$. Now, from the cumulant generating function, the central moments can be obtained, by using the expression,

$$\frac{d^n}{dt} K_S(t) \bigg|_{t=0} = E((S - E(S))^n),$$

where $K_S(t) = \log M_S(t) = r \log \frac{p}{1-(1-p)M_X(t)}$.

If N has any other model for the counting distribution, similar compound models for S can be obtained. For example if the distribution of N is binomial, then S has a compound binomial distribution.

2.3 Classes of discrete probability distributions

In the following we will introduce two important families of counting distributions that will be useful in the next section to calculate aggregate claims distribution.

2.3.1 The $(a, b, 0)$ class of distributions

Definition 2.2 *Let* $\Pr(N = n) = p_n$ *the probability mass function of a discrete random variable. The distribution is said to belong to the $(a, b, 0)$ class if there exist constants a and b such that*

$$\Pr(N = n) = \left(a + \frac{b}{n}\right) p_{n-1}, \quad n = 1, 2, \ldots$$

In the following theorem without proof, we provide the members of this important class of probability distributions:

Theorem 2.3 *The only non-trivial members of the $(a, b, 0)$ class are the Poisson, binomial, geometric and negative binomial distributions.*

Example 2.2 *Show that the Poisson distribution belongs to the $(a, b, 0)$ class.*

Solution: We prove the result by induction:
For $n = 1$, we have

$$p_1 = e^{-\lambda}\left(0 + \frac{\lambda}{1}\right) = e^{-\lambda}\lambda = \lambda p_0,$$

for $n = 2$, we have

$$p_1 = e^{-\lambda}\lambda\left(0 + \frac{\lambda}{2}\right) = \frac{\lambda}{2}p_1,$$

we assume the result is valid for $n - 1$, then

$$p_n = p_{n-1}\left(0 + \frac{\lambda}{n}\right) = e^{-\lambda}\frac{\lambda^{n-1}}{n-1!}\frac{\lambda}{n} = e^{-\lambda}\frac{\lambda^n}{n!}.$$

Then, $a = 0$, $b = \lambda$ and $p_0 = e^{-\lambda}$. □

2.3.2 The $(a, b, 1)$ class of distributions

Definition 2.3 *Let* $\Pr(N = n) = p_n$ *the probability mass function of a discrete random variable. The distribution is said to belong to the $(a, b, 1)$ class if there exist constants a and b such that*

$$\Pr(N = n) = \left(a + \frac{b}{n}\right) p_{n-1}, \quad n = 2, 3, \ldots$$

2.3. CLASSES OF DISCRETE PROBABILITY DISTRIBUTIONS

The difference from the $(a, b, 0)$ class is that the recursion starts from p_1 instead of p_0. The value of p_0 may or may not be zero. The members of the class $(a, b, 0)$ also belong to the $(a, b, 1)$ class.

Result 2.1 *The zero-truncated versions of the $(a, b, 0)$ class belong to the $(a, b, 1)$ class.*

Proof: This result can be easily achieved by removing the probability at 0, then re-scale the remaining probabilities. ∎

Result 2.2 *The zero-modifying versions of the $(a, b, 0)$ class belong to the $(a, b, 1)$ class.*

Proof: This result can be easily achieved by removing the probability at 0, then re-scale the remaining probabilities. ∎

Result 2.3 *The logarithmic distribution with probability mass function given by*

$$\Pr(N = n) = -\frac{1}{\log(1-\theta)} \cdot \frac{\theta^n}{n}, \quad n = 1, 2, \ldots,$$

where $0 < \theta < 1$, is a member of the the $(a, b, 1)$ class.

Proof: The result is shown by taking $a = \theta$, $b = -\theta$ and $p_1 = -\frac{\theta}{\log(1-\theta)}$. ∎

Example 2.3 *In this example, we consider the number of automobile insurance claims (see Willmot, 1987 and Gómez-Déniz and Calderín-Ojeda, 2015b). The number of claims and observed frequency are shown in Table 2.1 (first and second columns). We are fitting to this dataset the distributions in the $(a, b, 0)$ class and estimate their parameters by using the method of maximum likelihood. We are also computing the expected frequency for each model and compare the fit to data for each model using different measures of model validation.*

Solution: The results presented in this table are based on the maximum likelihood estimates of the parameters of those distributions.

Based on the results obtained the negative binomial distribution provides the best fit to data. This result is confirmed by the maximum of the log-likelihood function and the χ^2 test statistics, i.e., lowest value for this model. However, it is noted that both negative binomial and geometric distributions are not rejected, when the χ^2 test is used, at the usual nominal levels as judged by their p-values. The second best fit to data is given by the Geometric distribution in terms of the NLL. This is because that these two distributions are overdispersed (i.e., variance is larger than the mean). For this dataset, the sample mean is 0.1442 and the sample variance is 0.1639. Then the Poisson distribution, this distribution is not suitable to model this set of data since Poisson distribution is equidispersed. Finally the worst fit to the data is the Binomial distribution, the reason behind this is that the binomial model is infradispersed or underdispersed. Note that we have fixed parameter $m = 6$ to estimate p. □

Table 2.1: Observed and fitted automobile insurance claims for models in the $(a, b, 0)$ class.

Number of claims	Observed	Poisson $\mathcal{P}(\lambda)$	Geometric $\mathcal{G}e(p)$	Binomial $\mathcal{B}i(6, p)$	Negative Binomial $\mathcal{NB}(r, p)$
0	20592	20420.9	20615.8	20385.0	20596.8
1	2651	2945.1	2598.5	3012.3	2631.0
2	297	212.37	327.52	185.47	318.37
3	41	10.21	41.28	6.09	37.81
4	7	0.37	5.20	0.11	4.45
5	0	0.01	0.66	< 0.00	0.52
6	1	< 0.00	0.08	< 0.00	0.06
Total	23589	23589	23589	23589	23589
Estimates					
$\hat{\lambda}, \hat{p}$		0.1442	0.8740	0.0240	0.8857
$\hat{r}$		—	—	—	1.1179
NLL		10297.8	10224.0	10348.0	10223.4
χ^2 statistic		204.09	4.648	407.70	3.6118
Degrees of freedom		2	3	2	2
p-value		< 0.0001	0.1995	< 0.0001	0.1643

2.4 A recursive expression for the aggregate claims distribution

When the claims number distribution belongs to the $(a, b, 0)$ class of distributions, the Panjer's recursion formula (Panjer, 1981) allows to derive a recursive calculation of the aggregate claims amount when the individuals' claims sizes are distributed on the non-negative integers. The advantage of this expression is that there is no need to compute convolutions in the calculation of the probability function of the aggregate claim amount. This is formalized below.

Let us consider the same assumptions as in Subsection 2.2.2 and the random variable $S = \sum_{i=1}^{N} X_i$, represents the aggregate claims amount over a fixed period of time. As the claims number distribution belongs to the $(a, b, 0)$ class of distributions, we have that

$$\Pr(N = n) = \left(a + \frac{b}{n}\right) p_{n-1}, \quad n = 1, 2, \ldots$$

Let X_1 be distributed on the non-negative integers with probability function $f_j = \Pr(X_1 = j)$. Then Panjer's recursion gives the distribution of S

$$f_S(x) = \frac{q}{1 - a f_0} \sum_{k=1}^{x} \left(a + \frac{bk}{x}\right) f_k f_S(x - k),$$

and the starting value

$$f_S(0) = P_N(f_0) = p_0 + \sum_{n=0}^{\infty} p_n f_0^n.$$

For the case that N follows a geometric distribution,

$$p_n = \Pr(N = n) = pq^n, \quad n = 0, 1, 2, \ldots,$$

belongs to the $(a, b, 0)$ class with $a = q$ and $b = 0$. Then, assuming that X_1 is distributed on the non-negative integers with probability function $f_j = \Pr(X_1 = j)$. Then Panjer's recursion for the distribution of S is given by

$$f_S(x) = \frac{q}{1 - q f_0} \sum_{k=1}^{x} f_k f_S(x - k),$$

and the starting value

$$f_S(0) = P_N(f_0) = \sum_{n=0}^{\infty} p (q f_0)^n = \frac{p}{1 - q f_0}.$$

Then for $y = 1, 2, 3, \ldots$,

$$\begin{aligned}
F_S(y) = \Pr(S \le y) &= \sum_{x=0}^{y} f_S(x) = f_S(0) + \sum_{x=1}^{y} \frac{q}{1-qf_0} \sum_{k=1}^{x} f_k f_S(x-k) \\
&= \frac{p}{1-qf_0} + \frac{q}{1-qf_0} \sum_{k=1}^{y} f_k \sum_{x=k}^{y} f_S(x-k) \\
&= \frac{p}{1-qf_0} + \frac{q}{1-qf_0} \sum_{k=1}^{y} f_k F_S(y-k).
\end{aligned}$$

Thus, the distribution function can be calculated recursively, starting from $F_S(0) = f_S(0)$.

Similarly, when the counting distribution belongs to the $(a, b, 1)$ class and the individual claims amount are distributed on the non-negative integers the procedure used above can be used to derive a recursion formula for the distribution of the aggregate claims amount. See for instance, Sundt and Vernic (2009).

Moreover, if the probability mass function of a distribution can be written in terms of a recursive formula, then the distribution of the aggregate claims amount can be obtained via an integral equation. In the following example, we are firstly deriving a recursive formula for the probability mass function of the $\mathcal{NBIG}$ (Gómez-Déniz et al., 2008) distribution with probability function given by (1.23), which will be denoted as $p_r(x)$. Next result gives this recursive expression.

Theorem 2.4 *The probability mass function of the $\mathcal{NBIG}$ distribution can be evaluated by the recursive formula*

$$p_r(x) = \frac{r+x-1}{x}\left[p_r(x-1) - \frac{r}{r+x-1}p_{r+1}(x-1)\right], \quad x = 1, 2, \ldots \quad (2.5)$$

Proof: For the negative binomial distribution with pmf

$$p(x|\lambda) = \binom{r+x-1}{x} e^{-\lambda r}(1-e^{-\lambda})^x, \quad x = 0, 1, \ldots$$

we have the simple recursion

$$\frac{p(x|\lambda)}{p(x-1|\lambda)} = \frac{r+x-1}{x}(1-e^{-\lambda}), \quad x = 1, 2, \ldots \quad (2.6)$$

Using the definition of a $\mathcal{NBIG}$ distribution and (2.6) we get

$$\begin{aligned}
p_r(x) &= \int_0^\infty p(x|\lambda) f(\lambda)\, d\lambda \\
&= \int_0^\infty \frac{r+x-1}{x}(1-e^{-\lambda}) p(x-1|\lambda) f(\lambda)\, d\lambda \\
&= \frac{r+x-1}{x}\left[p_r(x-1) - \int_0^\infty e^{-\lambda} p(x-1|\lambda) f(\lambda)\, d\lambda\right].
\end{aligned}$$

2.4. A RECURSIVE EXPRESSION FOR THE AGGREGATE CLAIMS DISTRIBUTION

Now,

$$\int_0^\infty e^{-\lambda} p(x-1|\lambda) f(\lambda)\, d\lambda = \binom{r+x-2}{x-1}$$
$$\times \int_0^\infty e^{-\lambda(r+1)}(1-e^{-\lambda})^{x-1} f(\lambda)\, d\lambda$$
$$= \frac{r}{r+x-1}\binom{r+x-1}{x-1}$$
$$\times \int_0^\infty e^{-\lambda(r+1)}(1-e^{-\lambda})^{x-1} f(\lambda)\, d\lambda$$
$$= \frac{r}{r+x-1} p_{r+1}(x-1),$$

and therefore we obtain (2.5). ■

In the following example, we will show that if the claims sizes are absolutely continuous random variables with pdf $f(x)$ for $x > 0$, then the density of the aggregate claims amount can be obtained via an integral equation given that the number of claims follows a $\mathcal{NBIG}$ distribution.

Example 2.4 *If the claim sizes are absolutely continuous random variables with pdf $f(x)$ for $x > 0$, then the density $f_S(x;r)$ of the compound $\mathcal{NBIG}$ distribution satisfies the integral equation,*

$$f_S(x;r) = p_r(0) + \int_0^x \frac{ry+x-y}{x} f_S(x-y;r) f(y)\, dy$$
$$- \int_0^x \frac{ry}{x} f_S(x-y;r+1) f(y)\, dy. \qquad (2.7)$$

Solution: We have that the aggregated claims distribution is given by

$$f_S(x;r) = \sum_{k=0}^\infty p_r(k) f^{k*}(x) = p_r(0) f^{0*}(x) + \sum_{k=1}^\infty p_r(k) f^{k*}(x),$$

where f^{k*} denotes the k-fold convolution of $f(x)$. Now, using (2.5) we have that

$$p_r(k) = \left(\frac{r-1}{k}+1\right) p_r(k-1) - \frac{r}{k} p_{r+1}(k-1), \quad k=1,2,\ldots$$

Then,

$$\sum_{k=1}^\infty p_r(k) f^{k*}(x) = \sum_{k=1}^\infty \frac{r-1}{k} p_r(k-1) f^{k*}(x) + \sum_{k=1}^\infty p_r(k-1) f^{k*}(x)$$
$$- \sum_{k=1}^\infty \frac{r}{k} p_{r+1}(k-1) f^{k*}(x).$$

Now, after some straightforward calculations and using the identities:

$$f^{k*}(x) = \int_0^x f^{(k-1)*}(x-y)f(y)\,dy, \quad k=1,2,\ldots$$

$$\frac{f^{k*}(x)}{k} = \int_0^x \frac{y}{x} f^{(k-1)*}(x-y)f(y)\,dy, \quad k=1,2,\ldots$$

we obtain the result.

Integral equation (2.7) must be solved numerically. There are several implementations and algorithms for solving the Volterra integral equation of the second kind, however, they need to be modified for them to be used in (2.7). □

2.5 Premium calculation principles

A premium is the payment that an insured makes for total or partial cover against a risk. More formally, the premium can be defined as the amount that the policy holder pays for insurance coverage (see Dickson, 2017). In this section, we provide and examines several mathematical methods to calculate premiums.

Let us suppose that the claims from a risk are distributed as a random variable X. We have the following definition:

Definition 2.4 *Let the functional expression $\mathcal{P}_X$ represents the premium to cover the risk X, then $\mathcal{P}_X = \psi(X)$, where $\psi(\cdot)$ is some function. We say that $\psi(\cdot)$ is a premium principle.*

2.5.1 Examples

We will consider some examples of premium principles.

Pure premium principle

In this case, the premium equals the expectation of the risk, therefore there is no loading for profit or against adverse claims experience, so it is not used in practice. Then, we have that $\mathcal{P}_X = E(X)$.

Result 2.4 $\mathcal{P}_X = E(X)$ *minimises the second order moment about $\mathcal{P}$.*

Proof: Differentiating with respect to $\mathcal{P}$ the expression $E((X-\mathcal{P})^2)$, we have that

$$\frac{\partial}{\partial \mathcal{P}} E((X-\mathcal{P})^2) = 2E(X-\mathcal{P})).$$

Now setting this expression equal to 0, we have the result using the linear property of the expectation. ■

2.5. PREMIUM CALCULATION PRINCIPLES

Expected value premium principle

In this example, the premium is calculated as follows $\mathcal{P}_X = (1+\theta)E(X)$, where $\theta > 0$. The quantity θ is known as the premium loading factor. The loading in the premium is $\theta E(X)$.

Note that this premium principle assigns the same premium to risks with the same mean and different variance regardless how large the tail of the distribution is. For example, if X is exponentially distributed with mean 1 and Y follows a Lomax distribution with shape parameter 2 and scale parameter 1, then $E(Y) = 1$. We have that $var(X) = 1$ and $var(Y)$ does not exist. We expect that $\mathcal{P}_Y$ should exceed $\mathcal{P}_X$, but both premiums are equal under the expected value premium principle.

Variance/Standard deviation premium principle

In the variance premium principle, the premium is calculated according to the expression:
$$\mathcal{P}_X = E(X) + \beta var(X)$$
where $\beta > 0$ is the loading factor. As this premium principle is not a scaled invariant, sometimes it is preferable to work with the standard deviation premium principle which is given by
$$\mathcal{P}_X = E(X) + \beta\sqrt{var(X)}.$$

Esscher premium priniciple

Let X be a continuous random variable defined on $(0, \infty)$ with density function $f(\cdot)$, cdf $F(\cdot)$ and mgf given by $M_X(t) = \int_0^\infty e^{tx} f(x)\, dx$. Let

$$g(x) = \frac{e^{tx} f(x)}{M_X(t)}, \qquad \text{with } t > 0. \tag{2.8}$$

Then $g(\cdot)$ is a proper density function. Let Y have distribution with distribution function
$$G(x) = \frac{1}{M_X(t)} \int_0^x e^{ty} f(y)\, dy.$$
$G(\cdot)$ is known as the Esscher Transform of $F(\cdot)$ with parameter t.

Example 2.5 *Find the mgf of the Esscher transform Y.*

Solution: For any $t > 0$ and $h > 0$, we have that

$$M_Y(t) = \frac{1}{M_X(t)} \int_0^x e^{hx} e^{tx} f(y)\, dy = \frac{M_X(t+h)}{M_X(t)}. \tag{2.9}$$

□

Obviously, if X is a discrete random variable with probability mass function $\Pr(X = x)$ with $x = 0, 1, 2, \ldots$, and the mgf exists and it is given by M_X, then

$$g(x) = \frac{e^{tx} f(x)}{M_X(t)}, \quad x = 0, 1, 2, \ldots, t > 0,$$

is a proper probability mass function. Then, for a random variable Y following this distribution, the mgf of Y is given by (2.9)

Note that the pdf (2.8) can be rewritten as

$$g(x) = w(x) f(x)$$

with

$$w(x) = e^{tx}/M_X(t),$$

as a weight function, then for a given value of $t > 0$, we have that $w(x)$ is an increasing function of x, that is, greater weight are given to tail probabilities for the random variable Y with density $g(x)$.

Result 2.5 *Let X and Y be two continuous random variables with densities given by $f(x)$ and $g(x)$ and whose mgf exist, then $f(x) > g(x)$ for $x < x_0$ and $g(x) > f(x)$, $x > x_0$.*

Proof: It is straightforward since $w(x)$ is an increasing function of x and $w(0) = \frac{1}{M_X(t)} < 1$ and $w(x_0) = 1$ where $x_0 = \frac{\log M_X(t)}{t}$. ∎

Result 2.6 *Let X and Y be two continuous random variables with densities given by $f(x)$ and $g(x)$, then X is riskier than Y.*

Proof: For any $z > x_0$, it is satisfied that

$$\bar{G}(z) = \int_z^\infty g(x)\, dx > \bar{F}(z) = \int_z^\infty f(x)\, dx.$$

∎

Definition 2.5 *The Esscher premium (Kamp, 1998) of X is defined as*

$$\mathcal{P}_X = E(Y) = \frac{E(e^{tX} X)}{E(e^{tX})}, \quad t > 0.$$

Therefore, we can obtain the Esscher premium of X by calculating the pure premium of Y.

2.5. PREMIUM CALCULATION PRINCIPLES

The Esscher premium satisfies the following properties:

(i) $\mathcal{P}_X \geq E(X)$ for $t > 0$ and $\mathcal{P}_X = E(X)$ for $t = 0$.

(ii) $\mathcal{P}_X = \dfrac{M'_X(t)}{M_X(t)}$.

(iii) $\mathcal{P}_X$ is increasing in t.

(iv) Let $\bar{\mathcal{P}}_X$ be such that $E(e^{tX}(X - \bar{\mathcal{P}}_X)^2)$ is minimised. Then,
$$\bar{\mathcal{P}}_X = \frac{E(Xe^{tX})}{E(e^{tX})}.$$

2.5.2 Properties of premium calculation principles

Many authors have suggested various requirements that any premium calculation principle should satisfy (see for example, Kaas et al., 2008, Dickson, 2017, among others). A list of requirements that a premium calculation principle should fulfil are given below:

- **No-rip off:**
  $\mathcal{P}_X \leq \max\{X\}$, for all random variables. It is useless to keep more capital than the maximal loss value.

- **Non-negative loading:**
  $\mathcal{P}_X \geq E(X)$, for all random variables. The minimal capital must exceed the expected loss, otherwise ruin becomes certain.

- **Translativity:**
  $\mathcal{P}_{X+c} = \mathcal{P}_X + c$ for all random variables and each constant c. Any increase in the liability by a deterministic amount c should result in the same increase in the capital. This implies that $\mathcal{P}_{X-\mathcal{P}_X} = 0$.

- **Constancy (or no unjustified loading):**
  $\mathcal{P}_c = c$. To deal with a loss of c, Insurer only needs to have a capital of the same amount at its disposal.

- **Additivity:**
  $\mathcal{P}_{X+Y} = \mathcal{P}_X + \mathcal{P}_Y$ for all independent random variables, X and Y can be reduced by diversification.

- **Positive homogeneity:**
  $\mathcal{P}_{cX} = c\mathcal{P}_X$ for all random variables and any constant c. Independence with respect to the monetary units used.

All the aforementioned premium principles satisfy the **Non-negative loading** property. However, the **No rip-off** property is only verified by the pure and Esscher premium principles.

Example 2.6 *Show that the Pure and Esscher premium principles satisfy the No rip-off property.*

Solution: By denoting as x_m, the maximum claim amount, then for the pure premium principle we have that $X \leq x_m$, then $\mathcal{P}_X = E(X) \leq x_m$. For the Esscher premium principle This property holds as

$$\mathcal{P}_X = \frac{E(Xe^{hX})}{E(e^{hX})} \leq \frac{E(x_m e^{tx_m})]}{E(e^{tx_m})} = \frac{x_m e^{tx_m}}{e^{tx_m}} = x_m.$$

□

In the following example we illustrate that that the expected value premium principle does not verify neither the translativity nor constancy properties.

Example 2.7 *Show that the expected value premium principle does not satisfy the neither translativity property nor the no unjustified loading.*

Solution: Let $Y = X + c$, then

$$\mathcal{P}_Y = (1+\theta)E(X+c) = (1+\theta)(E(X)+c) = \mathcal{P}_X + c + c\theta > \mathcal{P}_X + c$$

when $c > 0$, then the translativity is not verified. Similarly by denoting $X = c$, then

$$\mathcal{P}_X = (1+\theta)E(c) = (1+\theta)c \neq c,$$

and the constancy property is not verified either. □

All the premium principles confirm the **additivity** property except for the standard deviation premium principle.

Example 2.8 *Show that the standard premium principle does not satisfy the additivity property.*

Solution: Let X and Y be two independent random variables, then

$$\begin{aligned}\mathcal{P}_{X+Y} &= E(X+Y) + \alpha\sqrt{var(X+Y)} = E(X) + E(Y) + \alpha\sqrt{var(X+Y)} \\ &\neq E(X) + E(Y) + \alpha\sqrt{var(X) + var(Y))} = \mathcal{P}_X + \mathcal{P}_Y.\end{aligned}$$

and the property is not verified. □

It is simple to observe that pure, expected value and standard deviation premium principles verify the **positive homogeneity** property.

Example 2.9 *Show that the variance premium principle does not satisfy the positive homogeneity property.*

Solution: Let $Y = cX$, then

$$\begin{aligned}\mathcal{P}_Y &= E(cX) + \alpha\, var(cX) = cE(X) + \alpha c^2 var(X) \neq cE(X) + \alpha c\, var(X) \\ &= c\mathcal{P}_X,\end{aligned}$$

when $c > 0$, then the property is not verified. □

2.6 Risk measures

Risk measures (see for example, Denuit et al., 2005, Dhaene et al., 2006 Kaas et al., 2008, among others) are used for determining provisions and capital requirements to avoid insolvency. Since risks are modelled as non-negative random variables, measuring risk is equivalent to establishing a correspondence between the space of random variables and $\mathbb{R}^+$. We will focus on risk measures that measure the upper tails of distribution functions.

Definition 2.6 *Let the non-negative random variable X represent the amount of a loss (profit) following a distribution. A risk measure $\mathcal{P}_X = \mathcal{H}(X)$ is a functional which maps $X \in \mathcal{X}$ to a positive real number., i.e.,*

$$\mathcal{H}: \mathcal{X} \longrightarrow \mathbb{R}^+.$$

Here $\mathcal{H}(X)$ represents the extra cash to be added to X to make it acceptable to an external or internal risk controller. It is the risk capital of the portfolio.

Some of the applications of risk measures include:

- Premium calculation principles are some kinds of risk measures which are to determine premium charged for a loss amount X;
- Setting provisions and capital requirements to ensure solvency;
- pricing insurance and financial products;
- Ordering of risks.
- Comparing reinsurance contracts.

In the following we define some important risk measures.

2.6.1 Value at Risk (VaR)

Consider a risk X with a cdf $F_X(\cdot)$ and a probability level $\alpha \in (0,1)$.

Definition 2.7 *The value-at-risk (VaR) at level α is defined as the α-quantile of $F_X(\cdot)$ and it will be denoted as*

$$Q_\alpha(X) = \inf\{x \in \mathbb{R} : F_X(x) \geq \alpha\}, \quad \alpha \in (0,1),$$

where $F_X(x) = \Pr(X \leq x)$.

If $F_X(\cdot)$ is a continuous distribution function, the α-quantile of $F_X(\cdot)$ is defined to be a real number such that $Q_\alpha(X) = F_X^{-1}(\alpha)$ with $\Pr(X \leq Q_\alpha(X)) = \alpha$.

In general, the α-quantile of $F(\cdot)$ is defined to be a real number such that

$$Q_\alpha(X) = \min\{x : F_X(x) \geq \alpha\} = \min\{x : \bar{F}_X(x) \leq 1 - \alpha\},$$

with $\alpha \in (0,1)$. The α-quantile function $Q_\alpha(X)$ is a non-decreasing and left-continuous function of α.

We will often use the following equivalence relation which holds for all $x \in \mathbb{R}$ and $0 \leq \alpha \leq 1$:
$$Q_\alpha(X) \leq x \iff \alpha \leq F_X(x).$$

Note that the equivalence holds with equalities if F_X is continuous at this particular x.

Now, we state the following result without proof which is useful to calculate the VaR for transformation of random variables. The proof can be found in Dhaene et al. (2002):

Result 2.7 *Let X be a real-valued random variable, and $0 < \alpha < 1$. For any non-decreasing and left continuous function $g(\cdot)$, it holds that*
$$Q_\alpha(g(X)) = g(Q_\alpha(X)).$$

Example 2.10 *Consider a random variable X that is normally distributed with mean μ and variance σ^2. Calculate the value at risk of the random variable X.*

Solution: As $X = \mu + \sigma Z$ with $Z \sim N(0,1)$, then, given $\alpha \in (0,1)$, the VaR for the standard normal distribution is given by $Q_\alpha(Z) = \Phi^{-1}(\alpha)$.

As the function $g(z) = \mu + \sigma z$ is non decreasing and left continuous, then the value-at-risk of the normal distribution with mean μ and variance σ^2 is
$$Q_\alpha(X) = Q_\alpha(g(Z)) = g(Q_\alpha(Z)) = \mu + \sigma Q_\alpha(Z) = \mu + \sigma \Phi^{-1}(\alpha).$$
$\square$

Example 2.11 *Derive the value at risk of the lognormal distribution.*

Solution: The value at risk of the lognormal distribution can be obtained via Result 2.7 and taking into account that $\log X \sim \mathcal{N}(\mu, \sigma^2)$. Then, we have
$$Q_\alpha(X) = \exp[\mu + \sigma \Phi^{-1}(q)], \quad 0 < \alpha < 1,$$
where $\Phi^{-1}(z)$ is the quantile function of the standard normal distribution. $\square$

The following result shows that the integral of the value at risk alongside all the values of α coincides with the expectation of the random variable X.

Result 2.8 *It is verified that $E(X) = \int_0^1 Q_p(X)\, dp$.*

Proof: For any real-valued random variable X with cdf $F_X(\cdot)$. As $F_X^{-1}(U)$ has the same distribution than X when $U \sim (0,1)$. Then, by assuming that $E(X) < \infty$, we have that
$$E(X) = E(F_X^{-1}(U)) = \int_0^1 F_X^{-1}(u)\, du = \int_0^1 Q_p(X)\, dp.$$
■

2.6. RISK MEASURES

2.6.2 Tail Value at Risk (TVaR)

A single quantile risk measure of a predetermined level α does not give any information about the thickness of the upper tail of the distribution function from $Q_\alpha(X)$. For example, a regulator is not only concerned with the frequency of default, but also about the severity of default. In this case, it should be interesting to quantify the expected value of the loss given that an event outside a given probability level has occurred. For this, the practitioner often uses a different measure, which is called the tail value-at-risk (TVaR) (see for example, Acerbi and Tasche, 2002, Dhaene et al., 2006, Sarabia and Calderín-Ojeda, 2018 among others) and defined next.

Definition 2.8 *Given a risk X and a probability level α, the corresponding TVaR, denoted by $TVaR_\alpha(X)$ is defined by*

$$TVaR_\alpha(X) = \frac{1}{1-\alpha} \int_\alpha^1 Q_p(X)\,dp, \quad \alpha \in (0,1).$$

It is the arithmetic average of the quantiles of X, from α on.

Note that $\text{TVaR}_\alpha(X) \geq Q_\alpha(X)$ and it is also a non-decreasing function in α.

Result 2.9 *The $TVaR_\alpha(X)$ is a non-decreasing function in α.*

Proof: As $\text{TVaR}_\alpha(X) = \frac{1}{1-\alpha}\left(E(X) - \int_0^\alpha Q_p(X)\,dp\right)$, then

$$\frac{d}{d\alpha}\text{TVaR}_\alpha(X) = \frac{1}{1-\alpha}\text{TVaR}_\alpha(X) - \frac{Q_\alpha(X)}{1-\alpha}.$$

Now as $Q_\alpha(X)$ is non-decreasing in α, we have

$$\text{TVaR}_\alpha(X) = \frac{1}{1-\alpha}\int_\alpha^1 Q_p(X)\,dp \geq Q_\alpha(X),$$

which provides

$$\frac{d}{d\alpha}\text{TVaR}_\alpha(X) \geq 0.$$

Then,

$$\text{TVaR}_\alpha(X) \geq \text{TVaR}_0(X) \geq E(X).$$

Therefore, TVaR induces a non-negative loading whatever the probability level α. ∎

Example 2.12 *Derive the analytical expression of the Tail Value at Risk (TVaR) of the exponential distribution with pdf as in (1.12).*

Solution: By inverting the distribution function associated to (1.12), it is obtained that the value at risk:

$$Q_\alpha(X) = -\sigma \log(1-\alpha), \quad 0 < \alpha < 1.$$

Also, the tail value at risk is,

$$\text{TVaR}_\alpha(X) = \sigma[1 - \log(1-\alpha)], \quad 0 < \alpha < 1,$$

since we have integrated the VaR between α and 1

$$\begin{aligned}\text{TVaR}_\alpha(X) &= \frac{1}{1-\alpha} \int_\alpha^1 [-\sigma \log(1-p)] \, dp \\ &= \sigma[1 - \log(1-\alpha)].\end{aligned}$$

$\square$

2.6.3 Conditional Tail Expectation (CTE) and Expected Shortall (ES)

If X denotes the aggregate claims of an insurance portfolio over a given reference period, we could define "bad times" for this portfolio as those where X takes a value in the interval

$$[Q_\alpha(X), \text{TVaR}_\alpha(X)].$$

The aggregate claims exceed the threshold $Q_\alpha(X)$ but not use up all available capital. This leads us to the following definition:

Definition 2.9 *The conditional tail expectation at level α (see for example, Acerbi and Tasche, 2002, Denuit et al., 2005 Sarabia and Calderín-Ojeda, 2018, among others) is the conditional expected loss given that the loss exceeds its VaR:*

$$CTE_\alpha(X) = E(X|X > Q_\alpha(X)).$$

In other words, the conditional tail expectation at level α is equal to the mean of top $(1-\alpha)\%$ losses. It provides a cushion against the mean value of losses exceeding the critical threshold $Q_\alpha(X)$.

As the VaR at a fixed level only gives local information about the underlying distribution, a possible way to avoid this limitation is to consider the so-called expected shortfall over some quantile.

Definition 2.10 *Let us consider a risk X, then the **stop-loss premium** (see Kaas et al., 2008) with retention $c > 0$, it is defined as*

$$E[(X-c)_+] = \int_c^\infty (x-c) f_X(x) \, dx,$$

where $f_X(\cdot)$ is the density of the random variable X.

2.6. RISK MEASURES

Definition 2.11 *The expected shortfall (ES) at level α is the **stop-loss premium** with retention $Q_\alpha(X)$. It is defined as*

$$ES_\alpha(X) = E[(X - Q_\alpha(X))_+].$$

This risk measure can be interpreted as the expected value of the shortfall in case the capital is set equal to $Q_\alpha(X) - P$, where P is the provision for this portfolio.

The relationship between the different risk measures is given in the following result:

Result 2.10 *For $\alpha \in (0,1)$, it is verified that*

$$TVaR_\alpha(X) = Q_\alpha(X) + \frac{1}{1-\alpha} ES_\alpha(X) \qquad (2.10)$$

$$CTE_\alpha(X) = Q_\alpha(X) + \frac{1}{1 - F_X(Q_\alpha(X))} ES_\alpha(X) \qquad (2.11)$$

$$CTE_\alpha(X) = TVaR_{F_X(Q_\alpha(X))}(X). \qquad (2.12)$$

Note that $F_X(\cdot)$ is continuous then

$$CTE_\alpha(X) = TVaR_\alpha(X)$$

Proof: We have

$$\begin{aligned} ES_\alpha(X) &= \int_0^1 (Q_p(X) - Q_\alpha(X))_+ \, dp \\ &= \int_\alpha^1 Q_p(X) \, dp - Q_\alpha(X)(1-\alpha) \\ &= (1-\alpha)\text{TVaR}_\alpha(X) - Q_\alpha(X)(1-\alpha), \end{aligned}$$

and then, the result (2.10) follows immediately.

$$\begin{aligned} ES_\alpha(X) &= E[(X - Q_\alpha(X))|X > Q_\alpha(X)]\bar{F}_X(Q_\alpha(X)) \\ &= E(X|X > Q_\alpha)\bar{F}_X(Q_\alpha) - E(Q_\alpha|X > Q_\alpha)\bar{F}_X(Q_\alpha) \\ &= CTE_\alpha(X)\bar{F}_X(Q_\alpha(X)) - Q_\alpha(X)\bar{F}_X(Q_\alpha(X)), \end{aligned}$$

and then, the result (2.11) follows. Finally, (2.12) follows straightforward from (2.10) and (2.11). ∎

Example 2.13 *Consider a random variable X is exponentially distributed with mean $1/\theta$. Show that*

$$CTE_\alpha(X) = TVaR_\alpha(X)$$

when $\alpha = 0.90$ and $\theta = 0.5$.

Solution: The VaR for the exponential distribution is given by

$$Q_\alpha(X) = -\frac{\log(1-\alpha)}{\theta}.$$

The TVaR is given by

$$\text{TVaR}_\alpha(X) = \frac{1}{1-\alpha}\int_\alpha^1 -\frac{\log(1-p)}{\theta}\,dp = \frac{1-\log(1-\alpha)}{\theta} = 6.605.$$

Alternatively,

$$\begin{aligned}
\text{CTE}_\alpha(X) &= \text{TVaR}_\alpha(X) = E(X|X>Q_\alpha(X)) \\
&= \frac{1}{1-F(Q_\alpha(X))}\int_{Q_\alpha(X)}^\infty xf(x)\,dx \\
&= \frac{1}{1-\alpha}\frac{\exp\{-\theta Q_\alpha(X)\}(1+Q_\alpha(X)\theta)}{\theta} \\
&= \frac{1-\log(1-\alpha)}{\theta} = 6.605.
\end{aligned}$$

□

2.6.4 Properties of risk measures

Many authors (see for example, , Denuit et al., 2005, Kaas et al., 2008 among others) have suggested various requirements that any risk measure should satisfy. A list of requirements that a risk measure should fulfil are given below:

- **No-rip off:**
  $\mathcal{H}_X \leq \max\{X\}$, for all random variables. It is useless to keep more capital than the maximal loss value.

- **Non-negative loading:**
  $\mathcal{H}_X \geq E(X)$, for all random variables. The minimal capital must exceed the expected loss, otherwise ruin becomes certain.

- **Translativity:**
  $\mathcal{H}_{X+c} = \mathcal{H}_X + c$ for all random variables and each constant c. Any increase in the liability by a deterministic amount c should result in the same increase in the capital. This implies that $\mathcal{H}_{X-\mathcal{H}_X} = 0$.

- **Constancy (or no unjustified loading):**
  $\mathcal{H}_c = c$. To deal with a loss of c Insurer only needs to have a capital of the same amount at its disposal.

- **Subadditivity:**
  $\mathcal{H}_{X+Y} \leq \mathcal{H}_X + \mathcal{H}_Y$ for all random variables X and Y. Subadditivity reflects the idea that risk can be reduced by diversification.

2.7. REINSURANCE

- **Positive homogeneity:**
  $\mathcal{H}_{cX} = c\mathcal{H}_X$ for all random variables and any constant c. Independence with respect to the monetary units used.

The value at risk (VaR) satisfies all the above mentioned properties except for the subadditivity property. A counter example to show this can be found in Kaas et al. (2008). This implies that that under the VaR the risk cannot in general be reduced by diversification. Moreover, the tail value at risk (TVaR) verifies all seven properties.

Definition 2.12 *A risk measure that is translative, positive homogeneous, subadditive and monotone is called coherent.*

2.7 Reinsurance

Reinsurance is the procedure by which insurers transfer part of their risk portfolios to other parties by some form of contract to reduce the probability of paying a large obligation due to a claim (see, for example, Kaas et al., 2008, Albrecher et al., 2017, Dickson, 2017 among others). The party that diversifies its insurance portfolio is known as the ceding company or insurer. The party that accepts a portion of the potential obligation in exchange for a share of the insurance premium is known as the reinsurer. Broadly speaking, reinsurance is known as insurance for insurers. Reinsurance allows insurers to remain solvent by recovering some or all amounts paid to claimants. Reinsurance reduces the net liability on individual risks and catastrophe protection from significant or multiple losses by paying to the reinsurer a reinsurance premium. This process provides the ceding companies the capacity to increase their underwriting possibilities in terms of the number and size of risks. This practice allows the insurer to reduce the risk, but the volume of business is also reduced in return. Mathematically, this is formalized as follows.

Let
$$S = \sum_{i=1}^{n} X_i$$
be the total claims from a homogenous portfolio of n policies, where $X_i, i = 1, 2, \ldots, n$, are assumed to be independent and identically distributed, representing claims from these n policies, respectively. A reinsurance arrangement is described by a measurable function
$$h : \mathbb{R}^n \longrightarrow \mathbb{R}^+,$$
where $h(x_1, x_2, \ldots, x_n)$ is the amount payable by the insurer if claims from these n policies are $x_1, x_2, \ldots, x_n$.

In order to define a proper reinsurance, we require that $h(\cdot)$ satisfies
$$0 \leq h(x_1, x_2, \ldots, x_n) \leq x_1 + x_2 + \cdots + x_n,$$
for all $x_1 \geq 0, x_2 \geq 0, \ldots, x_n \geq 0$.

Among all the reinsurance arrangements, firstly we differentiate two types of reinsurance: **individualized** type and **global** type. The individualized types of reinsurance are based on individual claims, mathematically,

$$h(x_1, x_2, \ldots, x_n) = \sum_{i=1}^{n} h_i(x_i),$$

where $0 \leq m_i(x_i) \leq x_i$ is the amount payable by the insurer if X_i takes value x_i and $\sum_{i=1}^{n} h_i(x_i)$ is the total claim payable by the insurer under this individualized types of reinsurance $h(\cdot)$. The global type reinsurance is based on the total amount of claims S, mathematically, $h(x_1, x_2, \ldots, x_n)$ is a function of $S = \sum_{i=1}^{n} x_i$ only.

These results can be extended for the case that N is a discrete random variable and the aggregate claim amount is given by $S = \sum_{i=1}^{N} X_i$.

2.7.1 Type of reinsurance

Let $S = \sum_{i=1}^{N} X_i$ be the aggregate claim in a year from an insurance portfolio, where all assumptions on S given in the collective risk model apply here. Let us define the following types of reinsurance

Definition 2.13 Proportional reinsurance:

Under proportional reinsurance, the reinsurer receives a certain proportion of all policy premiums sold by the insurer. For a claim, the reinsurer bears a the same proportion of the losses.

From a mathematical point of view

$$S_I = \sum_{i=1}^{N} a X_i = aS$$

denotes the claims amount retained by the insurer and by

$$S_R = \sum_{i=1}^{N} (1-a) X_i = (1-a)S,$$

the ceded amount to the reinsurer where $0 < a < 1$ is the retention level. Note that this type of reinsurance is of local and global type simultaneously.

Example 2.14 *Let X be the loss amount of a risk following an exponential distribution with mean $1/\lambda$. Consider the following reinsurance contract: the insurer pays $Y = \alpha X$ and the reinsurer pays $Z_1 = (1-\alpha)X$ with $0 < \alpha < 1$. Find the probability distribution of Y and Z.*

Solution: As X follows and exponential distribution with mean $1/\lambda$, its cdf is given by

$$F_X(x) = 1 - e^{-\lambda x},$$

2.7. REINSURANCE

then, the probability function of the amount paid by the insurer is given by

$$F_Y(x) = \Pr(Y \leq x) = \Pr(\alpha X \leq x) = \Pr\left(X \leq \frac{x}{\alpha}\right)$$
$$= F_Y\left(\frac{x}{\alpha}\right) = 1 - e^{-\frac{\lambda}{\alpha}x},$$

therefore Y follows an exponential distribution with mean α/λ.

Similarly, the probability function of the amount ceded to the reinsurer is given by

$$F_Z(x) = \Pr(Z \leq x) = \Pr((1-\alpha)X \leq x) = \Pr\left(X \leq \frac{x}{1-\alpha}\right)$$
$$= F_Z\left(\frac{x}{1-\alpha}\right) = 1 - e^{-\frac{\lambda}{1-\alpha}x},$$

therefore Z follows an exponential distribution with mean $(1-\alpha)/\lambda$. □

Definition 2.14 Excess of Loss reinsurance:
This is a type of local type reinsurance in which the reinsurer covers the losses exceeding the insurer's retention level M. That is if X is a random variable representing the size of a claim then,

i) *if $X \leq M$ the insurer pays the full claim and no reinsurance;*

ii) *if $X > M$, reinsurance is feasible and reinsurer pays $X - M$.*

Therefore the insurer pays $Y = \min(X, M)$ and the reinsurer pays $Z = \max(0, X - M) = (X - M)_+$.

Therefore if $S = \sum_{i=1}^{N} X_i$ represents the aggregate claim in a year from an insurance portfolio, we have that

$$S_I = \sum_{i=1}^{N} \min(X_i, M) = \sum_{i=1}^{N} Y_i$$

denotes the claims amount retained by the insurer and by

$$S_R = \sum_{i=1}^{N} \max(0, X_i - M) = \sum_{i=1}^{N} Z_i,$$

the ceded amount to the reinsurer where the retention level M is fixed.

Example 2.15 *Let X be the loss amount of a risk following an exponential distribution with mean $1/\lambda$. Consider the following reinsurance contract: the insurer pays $Y = \min(X, M)$ and the reinsurer pays $Z_1 = \max(0, X - M)$ with $M > 0$. Find the probability distribution and expectation of Y and Z.*

Solution: As X follows and exponential distribution with mean $1/\lambda$, its cdf is given by
$$F_X(x) = 1 - e^{-\lambda x},$$
then, the probability function of the amount paid by the insurer is given by
$$F_Y(x) = \begin{cases} F_X(x) & x \leq M, \\ 1 & x > M, \end{cases}$$
and then
$$F_Y(x) = \begin{cases} 1 - e^{-\lambda x} & x \leq M, \\ 1 & x > M. \end{cases}$$
Then, $E(Y)$ is calculated using the result
$$E(Y) = \int_0^M \bar{F}_Y(x)\,dx = \int_0^M e^{-\lambda x}\,dx = \frac{1 - e^{-\lambda M}}{\lambda}.$$

Similarly, the probability function of the amount ceded to the reinsurer is given by
$$F_Z(x) = \begin{cases} F_X(M) & x = 0, \\ F_X(M+x) & x > 0, \end{cases}$$
and then
$$F_Z(x) = \begin{cases} 1 - e^{-\lambda M} & x = 0, \\ 1 - e^{-\lambda(M+x)} & x > 0. \end{cases}$$
Then, $E(Z)$ is calculated using the result
$$E(Z) = \int_M^\infty \bar{F}_Y(x)\,dx = \int_M^\infty e^{-\lambda x}\,dx = \frac{1}{\lambda} e^{-\lambda M}.$$

$\square$

Definition 2.15 Stop-Loss reinsurance:
This is a type of global type reinsurance in which the reinsurer covers the losses exceeding the insurer's retention level d of the the aggregate claim in a year from an insurance portfolio S. That is,

i) if $S \leq d$ the insurer pays the full claim and no reinsurance;

ii) if $S > d$, reinsurance is feasible and reinsurer pays $S - d$.

Therefore the insurer pays $S_I = \min(S, d)$ and the reinsurer pays $S_R = \max(0, S - d) = (S - d)_+$.

2.8 Comparing risks

Ordering of risks is crucial in aspect in the actuarial profession. There are two reasons why a risk, that is, a non-negative random loss would be preferred to another: On the one hand, the other risk is larger, on the other hand, the other risk is riskier (having a thicker-tail). This second motivation is related to the random variables with thicker tails (thicker-tailed). The latter reason means that the probability of extreme values is larger, making a risk with equal mean less attractive because it is more spread, i.e., less predictable. Having thicker tails means having larger stop-loss premiums.

From the fact that a risk is smaller than another, one may deduce that it is also preferable in the mean-variance order. In this ordering, the practitioner prefers the risk with the smaller mean, and the variance is used as a tie-breaker. However, this ordering is inadequate for actuarial purposes, since it leads to wrong decisions.

2.8.1 Stochastic dominance

Definition 2.16 *Consider two random variables X and Y (representing two losses). X is said to precede Y in the stochastic dominance sense or X is said to be smaller than Y notation $X \leq_{st} Y$ (see for example, Shaked and Shanthikumar, 2007, Gómez-Déniz et al., 2013 among others) if and only if the distribution function of X always exceed that of Y:*

$$F_X(x) \geq F_Y(x), \quad -\infty < x < \infty.$$

Note that this expression is the same as

$$\bar{F}_X(x) = 1 - F_X(x) \leq \bar{F}_Y(x) = 1 - F_Y(x), \quad -\infty < x < \infty.$$

Alternatively, X is smaller than Y in the stochastic dominance sense if and only if their respective quantiles are ordered:

$$X \leq_{st} Y \iff Q_\alpha(X) \leq Q_\alpha(Y), \text{ for all } \alpha \in (0,1).$$

Theorem 2.5 *Two random variables X and Y satisfy $X \leq_{st} Y$ if, and only if, there exist two random variables $\hat{X}$ and $\hat{Y}$, defined on the same probability space, such that*

(i) $\hat{X} =_{st} {}^1 X$,

(ii) $\hat{Y} =_{st} Y$ and

(iii) $\Pr(\hat{X} \leq \hat{Y}) = 1$.

Proof: Obviously the three conditions imply that $X \leq_{st} Y$. Now, let $F_X(\cdot)$ and $F_Y(\cdot)$ be the cdf of X and Y respectively, and $F_X^{-1}(\cdot)$ and $F_Y^{-1}(\cdot)$ the corresponding right continuous inverse. Define, $\hat{X} = F_X^{-1}(U)$ and $\hat{Y} = F_Y^{-1}(U)$ where $U \sim U(0,1)$

---
[1]denotes equality in law.

random variable. Then, it is easy to see that the first two conditions are verified and by using (5) (or (6)) the third condition holds. ∎

In the following result we will show that two random variables whose probability density functions cross once are stochastically ordered.

Result 2.11 Let $X \sim F_X(\cdot)$ and $Y \sim F_Y(\cdot)$, with $F'_X(x) = f_X(x)$ and $F'_Y(x) = f_Y(x)$, then a sufficient condition for stochastic order is that the densities satisfies that $f_X(x) \geq f_Y(x)$ for small x and the opposite for large x.

Proof: If $f_X(x) > f_Y(x), x < c$ and $f_X(x) < f_Y(x), x > c$, then $F_X(x) > F_Y(x)$ and consequently $\bar{F}_X(x) = 1 - F_X(x) < \bar{F}_Y(x) = 1 - F_Y(x)$; on the other hand, if $x \geq c$, then $\int_x^\infty f_X(z)\, dz \leq \int_x^\infty f_Y(z)\, dz$ and then $\bar{F}_X(x) = 1 - F_X(x) \leq \bar{F}_Y(x) = 1 - F_Y(x)$. ∎

Result 2.12 If $X \leq_{st} Y$ then $E(X) \leq E(Y)$.

Proof: As $\bar{F}_X(x) \leq \bar{F}_Y(x)$, $x \geq 0$, then $\int_0^\infty \bar{F}_X(x)\, dx \leq \int_0^\infty \bar{F}_Y(x)\, dx$ and the result is easily derived. ∎

Remark 2.2 The reciprocal does not hold: $E(X) \leq E(Y)$ is not sufficient to conclude that $X \leq_{st} Y$.

Example 2.16 Let $X \sim \mathcal{B}er(p)$ with $p = 1/2$ and let Y follows $\Pr(Y = c) = 1$ for a given c such that $1/2 < c < 1$, show that $X \not\leq_{st} Y$.

Solution: Clearly, $1/2 = E(X) \leq c = E(Y)$. The survival function of X is given by
$$\bar{F}_X(x) = \begin{cases} 1, & x < 0, \\ 1/2, & 0 \leq x < 1, \\ 0, & x \geq 1. \end{cases}$$

On the other hand the survival function of Y is
$$\bar{F}_Y(x) = \begin{cases} 1, & x < c, \\ 0, & x \geq c. \end{cases}$$

Therefore $X \not\leq_{st} Y$. □

Remark 2.3 The stochastic order $\leq_{st}$ can be considered as a generalization of the usual order $\leq$ for real numbers; i.e., given $\beta \leq \gamma \in \mathbb{R}$, we have that $\beta \leq_{st} \gamma$ (γ and β are viewed as degenerate random variables).

Example 2.17 The shifted exponential distribution with parameters $\theta > 0$ and $a \in \mathbb{R}$ is defined as
$$F_X(x) = \begin{cases} 1 - \exp(-\theta(x - a)) & \text{if } x \geq a, \\ 0 & \text{otherwise.} \end{cases}$$

2.8. COMPARING RISKS

(i) Illustrate that the exponential distribution is included as particular case.

(ii) For two random variables X and Y following the shifted exponential distribution, give the sufficient and necessary condition for $X \leq_{st} Y$.

Solution: Let X and Y be two shifted exponential random variables with respective parameters (θ, a) and (λ, b).

(i) The special case $a = 0$ corresponds to the $Exp(\theta)$.

(ii) We have that
$$X \leq_{st} Y \iff \theta \geq \lambda \text{ and } a \leq b.$$

$\square$

Remark 2.4 *In economics, the random variables represent fortunes, incomes, etc., so that $X \leq_{st} Y$ means that all the profit-seeking decision-makers prefer Y over X, since Y is larger than X. In actuarial science, however, the risks to be compared represent future random financial losses, so that $X \leq_{st} Y$ means that all the profit-seeking actuaries prefer X over Y, since the loss X is smaller than Y.*

Remark 2.5 *Note that it is also verified that $X \leq_{st} Y \iff -Y \leq_{st} -X$.*

2.8.2 Stochastic dominance and stop-loss premiums

We begin this subsection by providing the definition of *the stop-loss order* (see for example, Denuit et al., 2005, Dhaene et al., 2006, Kaas et al., 2008 among others).

Definition 2.17 *Consider two random variables X and Y (representing two losses). X is said to precede Y in the stop-loss order sense, notation $X \leq_{sl} Y$, if and only if X has lower stop-loss premium than Y:*
$$E[(X-d)_+] \leq E[(Y-d)_+], \quad 0 < d < \infty.$$

A random variable that is stop-loss larger than another risk with the same mean will be referred to as "more dangerous". Observe that stop-loss order, equality of the means $E(X) = E(Y)$ is not required.

In the following result it is shown the relationship between the stochastic order $X \leq_{st} Y$ and the difference between their respective stop-loss premiums given a level of retention d.

Result 2.13 *Given two random variables X and Y, $X \leq_{st} Y$ if, and only if, $f(d) = E[(Y-d)_+] - E[(X-d)_+]$ is non-increasing on $\mathbb{R}$.*

Proof: If $X \leq_{st} Y$ then $\bar{F}_Y - \bar{F}_X$ is non-negative so the result follows from

$$\int_d^\infty \bar{F}_Y(z)\,dz - \int_d^\infty \bar{F}_X(z)\,dz = \int_d^\infty (\bar{F}_Y(z) - \bar{F}_X(z))\,dz \text{ is non-increasing.}$$

Now, by assuming the non-increasingness of the difference

$$E[(Y-d)_+] - E[(X-d)_+].$$

Differentiating this expression with respect to d yields $\bar{F}_X(d) - \bar{F}_Y(d) \leq 0$, then $X \leq_{st} Y$ follows. ∎

When $X \leq_{st} Y$ holds, the difference between their respective stop-loss premiums decreases with the level of retention.

Now we provide the definition of convex order (see for example, Dhaene et al., 2006, Shaked and Shanthikumar, 2007, among others).

Definition 2.18 *Consider two random variables X and Y (representing two losses). X is said to precede Y in the convex order sense, notation $X \leq_{cx} Y$, if and only if $X \leq_{sl} Y$ and $E(X) = E(Y)$, assuming that the expectations exist.*

In the following result we illustrate the relationship between the stop-loss order and the tail value at risk.

Theorem 2.6 *For any random pair (X, Y) we have that X is smaller than Y in stop-loss order sense if and only if their respective TVaR's are ordered:*

$$X \leq_{sl} Y \iff TVaR_\alpha(X) \leq TVaR_\alpha(Y), \quad \text{for all } \alpha \in (0,1).$$

Proof: The proof of this result can be found in Dhaene et al. (2006). ∎

Example 2.18 *Show that given two random variables $X \sim Exp(\theta)$ and $Y \sim Exp(\lambda)$, then if $\theta \geq \lambda \implies X \leq_{sl} Y$.*

Solution: We have that

$$\text{TVaR}_\alpha(X) = \frac{1}{1-\alpha} \int_\alpha^1 -\frac{\log(1-p)}{\theta}\,dp = \frac{1-\log(1-\alpha)}{\theta}, \text{ and}$$

$$\text{TVaR}_\alpha(Y) = \frac{1}{1-\alpha} \int_\alpha^1 -\frac{\log(1-p)}{\lambda}\,dp = \frac{1-\log(1-\alpha)}{\lambda},$$

for $\alpha \in (0,1)$. As if $\theta \geq \lambda$ then $\text{TVaR}_\alpha(X) \leq \text{TVaR}_\alpha(Y)$ and therefore $X \leq_{sl} Y$. □

Remark 2.6 *It is verified that*

$$CTE_\alpha(X) \leq CTE_\alpha(Y) \Rightarrow X \leq_{sl} Y, \quad \text{for all } \alpha \in (0,1).$$

2.8. COMPARING RISKS

Proof: By using (2.12) in Result 2.10, then the identity

$$\text{TVaR}_{F_X(d)}(X) = \text{CTE}_{F_X(d)}(X)$$

holds for any d such that $0 < F_X(d) < 1$. Hence, the result follows using the reasoning of the first implication in the previous theorem. ∎

However, the other implication is not true in general. In fact, it is verified that

$$X \leq_{cx} Y \not\Rightarrow \text{CTE}_\alpha(X) \leq \text{CTE}_\alpha(Y), \quad \text{for all} \ \alpha \in (0,1).$$

This can be seen in the following counterexample.

Example 2.19 Let X and Y be two random variables where $Y \sim U(0,1)$ and X has cdf given by

$$F_X(x) = \begin{cases} x & 0 \leq x < 0.85, \\ 0.85 & 0.85 \leq x < 0.9, \\ 0.95 & 0.9 \leq x < 0.95, \\ x & 0.95 \leq x \leq 1. \end{cases}$$

Show that $X \leq_{cx} Y \not\Rightarrow CTE_\alpha(X) \leq CTE_\alpha(Y)$ but it is not verified that $CTE_\alpha(X) \leq CTE_\alpha(Y)$, for all $\alpha \in (0,1)$.

Solution: Clearly, $F_X(x) \leq F_Y(x)$ for $x < 0.9$, and $F_X(x) \geq F_Y(x)$ for $x \geq 0.9$. Also, $E(X) = E(Y) = 0.5$ and $X \leq_{sl} Y$, hence that $X \leq_{cx} Y$. However, it is easy to check that $\text{CTE}_{0.9}(X) > \text{CTE}_{0.9}(Y)$. □

Given two random variables X and Y denoting two losses and verifying $X \leq_{sl} Y$, although all risk averse decision makers prefer losing X, this is not the case in the premiums needed to compensate this loss:

Example 2.20 Consider the standard deviation principle

$$\mathcal{P}_X = E(X) + \alpha\sqrt{var(X)}, \quad \text{with} \ \alpha > 0.$$

If $X \sim Bin(1, 1/2)$ and $Y \equiv 1$, while $\alpha > 1$, show that $X \leq_{sl} Y$ but $\mathcal{P}_X \geq \mathcal{P}_Y$

Solution: First, it is simple to observe that $X \leq_{sl} Y$. Now, we have that

$$\mathcal{P}_X = \frac{1}{2} + \alpha\frac{1}{2} > 1 = \mathcal{P}_Y, \quad \text{with} \ \alpha > 1,$$

even though $\Pr(X \leq Y) = 1$. □

2.8.3 Stop-Loss order and Stop-Loss Reinsurance

We first show that two probability density functions that cross twice imply that their respective cumulative distribution functions cross once. This can be seen in the following lemma:

Lemma 2.1 *Let X and Y be two risks with $E(X) = E(Y)$ and $f_X(\cdot) \not\equiv f_Y(\cdot)$. If disjoint intervals $I_1 = [0, x_1)$, $I_2 = (x_1, x_2)$ and $I_3 = (x_2, \infty)$ exist and $I_1 \cup I_2 \cup I_3 = [0, \infty)$ and I_2 between I_1 and I_3, such that the densities of X and Y satisfy $f_X(x) \leq f_Y(x)$ both on I_1 and I_3, while $f_X(x) \geq f_Y(x)$ on I_2, then the cumulative distribution functions of X and Y, i.e., $F_X(\cdot)$ and $F_Y(\cdot)$, cross only once.*

Proof: Let us assume that $F_X(\cdot)$ and $F_Y(\cdot)$, then either $\bar{F}_X(x) < \bar{F}_Y(x)$ for all x or $\bar{F}_X(x) > \bar{F}_Y(x)$ for all x, it means that $E(X) < E(Y)$ or $E(X) > E(Y)$. Now we define
$$h(x) = \bar{F}_X(x) - \bar{F}_Y(x) = \int_x^\infty f_X(z) - f_Y(z)\,dz, \quad x \geq 0.$$

We have

(i) $h(0) = 1 - 1 = 0$;

(ii) $h(+\infty) = 0 - 0 = 0$;

(iii) $h(x) < 0$ for $x \in I_3$;

(iv) $h'(x) = f_Y(x) - f_X(x) > 0$ if $x \in I_1$, $h'(x) = f_Y(x) - f_X(x) < 0$ if $x \in I_2$ and $h'(x) = f_Y(x) - f_X(x) > 0$ if $x \in I_3$.

Then, there exists $\bar{x} \in I_2$, such that $h(\bar{x}) = 0$, and therefore $h(x) > 0$ and $\bar{F}_X(x) > \bar{F}_Y(x)$ for $0 < x < \bar{x}$ and $h(x) < 0$ and $\bar{F}_X(x) < \bar{F}_Y(x)$ for $x > \bar{x}$. ∎

Then, among the reinsurance contracts with the same expected value of the retained risk, stop-loss reinsurance gives the lowest possible variance.

Result 2.14 *Let us suppose that the random loss is X, and compare the cdf of the retained loss $Z = X - (X - d)_+$ under stop-loss reinsurance with another retained loss $Y = X - I(X)$ where $I(\cdot)$ is non-negative and $E(Y) = E(Z)$ then $Z \leq_{sl} Y$.*

Proof: As $I(\cdot)$ is non-negative, then it follows that $Y \leq X$ holds, and hence $F_Y(x) \geq F_X(x)$ for all $x > 0$. Further, $Z = \min\{X, d\}$, therefore $F_Z(x) = F_X(x)$ for all $x < d$, and $F_Z(x) = 1$ for $x \geq d$. Clearly, the cdfs of Z and Y cross exactly once, at d, and Y is more dangerous, then $Z \leq_{sl} Y$. ∎

Exercises

1. Show that the third central moment of total claims on a portfolio with n policies is given by
$$E(S - E(S))^3 = n(pE(S^3) - 3p^2 E(S^2) - 2p^3 \mu^3).$$

2.8. COMPARING RISKS

2. Prove that:

    a) The binomial distribution belongs to $(a, b, 0)$ class with $a = -\frac{p}{1-p}$, $b = (n+1)\frac{p}{1-p}$ and $p_0 = (1-p)^n$;

    b) the *negative binomial* to $(a, b, 0)$ class with $a = 1-p$, $b = (r-1)(1-p)$ and $p_0 = (1-p)^r$.

3. Let us consider the dataset given in Table 2.1. Fit the Poisson-inverse Gaussian distribution to this dataset by using the method of moments. Show that the estimates are $\hat{\mu} = 0.1442$ and $\hat{\gamma} = 0.1527$.

4. Let us assume that the number of claims follows a Poisson distribution with parameter 3 and the individual claims amount follows $fj = 0.7(0.3)^{j-1}$, for $j = 1, 2, \ldots$. Show that $f_S(0) = 0.0498$, $f_S(1) = 0.1046$ and $f_S(2) = 0.0011$.

5. Let us suppose that $X \sim N(\mu, \sigma^2)$. Show that under the Esscher premium principle is not satisfied $\mathcal{P}_{kX} \neq k\mathcal{P}_X$ where $k > 0$.

6. The aggregate claims from a risk have a compound Poisson distribution with Poisson parameter $\lambda = 5$ and individual claim amount are exponentially distributed with mean $\sigma = 15$. Show that the premium of this risk using the expected value premium principle with loading factor 0.2 is 90.

7. The aggregate claims from a risk is given by $S = \sum_{i=1}^{N} X_i$. It is known that $E(S) = \mu$ and $var(S) = \sigma^2$. The insurer has effected proportional reinsurer with retention level α with $0 < \alpha < 1$, where S_I represents the retained amount by the insurer and S_R, the ceded amount to the reinsurer. Show that $cov(S_I, S_R) = \alpha(1-\alpha)\sigma^2$.

8. Show that the tail value at risk (TVaR) satisfy the following results:

    a) $\text{TVaR}_\alpha(c) = c$, where c is a positive real number.
    b) $\text{TVaR}_\alpha(X + c) = \text{TVaR}_\alpha(X) + c$.
    c) $\text{TVaR}_\alpha(cX) = c\text{TVaR}_\alpha(X)$ where c is a positive real number.

9. Let X and Y be two random variables where $Y \sim U(0, 1)$ and X has cdf given by
$$F_X(x) = \begin{cases} x & 0 \leq x < 0.85, \\ 0.85 & 0.85 \leq x < 0.9, \\ 0.95 & 0.9 \leq x < 0.95, \\ x & 0.95 \leq x \leq 1. \end{cases}$$
Show that for $\alpha = 0.9$, $\text{CTE}_\alpha(X) = 0.975$ and $\text{CTE}_\alpha(Y) = 0.95$.

10. Let us consider two random variables X and Y that follow a Pareto type II or Lomax distribution with a probability density function given by $f_X(x) = \theta_1 \frac{\lambda^{\theta_1}}{(x+\lambda)^{\theta_1+1}}$ and $f_Y(x) = \theta_2 \frac{\lambda^{\theta_2}}{(x+\lambda)^{\theta_2+1}}$ with $x \geq 0$ where θ_i with $i = 1, 2$ are shape parameters and λ is the scale parameter with $E(X) = \frac{\lambda}{\theta_1-1}$ and $E(Y) = \frac{\lambda}{\theta_2-1}$, $\theta_1 > 1$ and $\theta_2 > 1$. If $1 < \theta_1 \leq \theta_2 < \infty$ and $0 < \lambda < \infty$. Show that

   a) $Y \leq_{st} X$ where $\leq_{st}$ denotes the usual stochastic order;

   b) $Y \leq_{sl} X$ where $\leq_{sl}$ denotes the stop-loss order.

Chapter 3

Statistical Distributions in Tourism

3.1 Introduction

Data collected in 2019 reveals that tourism accounted for 10 percent of global GDP, with a value of almost $9 trillion. These amounts suggest that the sector related to tourism is, for example, nearly three times larger than that of agriculture. However, it is a sector very susceptible to changes such as pandemics, wars, terrorist attacks, climate change, etc. On the other hand, contracting tourist trips has largely changed in recent decades with the possibility that the internet offers tourists to buy a plane or boat ticket, book a hotel or vacation home, rent a vehicle, etc. Added to this is that ecological tourism and cruise trips have experienced exponential growth in recent years.

Perhaps for these reasons, the number of scientific articles dealing with this sector has grown enormously in recent years, constituting an independent section in the extensive catalog of reports on economics.

An interesting body of literature in tourism statistics has focused on variables, such as the length of tourist stay at a vacation destination (a discrete random variable) and the expenditure (a continuous random variable).

Concerning the length of stay (LS) variable, this usually shows overdispersion (variance larger than the mean) and multimodality, specially bimodality. This is because most tourists hire packages of the duration of seven days or fourteen days. In addition, it is a censored variable since the support of that is concentrated in the set $\mathbb{N}^* = \{1, 2, \dots\}$.

On the other hand, one of the main characteristics of the empirical expenditure variable is the asymmetry and the long right tail that the empirical data offers in practice. Thus, few tourists spend a lot (at the origin, destination, or both, origin and destination together), and these data should not be disregarded.

Tourism managers must often evaluate the effectiveness of initiatives and programs such as marketing campaigns and economic development policies, taking into account various considerations, including sustainability targets and environmental issues (Edgell and Swanson, 2013). To define and plan the future direction of the industry in a tourist destination country, managers may base their analysis on variables such as tourists' length of stay, average expenditure or propensity to visit, and for this purpose, awareness of the joint data distribution and dynamics of the variables in question is necessary.

Focusing on the knowledge of the joint data distribution, policymakers can determine the impact of public policies, identifying the degree of dependence of relevant policy variables and the relationships between them. To do so, the analyst must take into account the level of dependence (high-low, positive-negative, symmetric-asymmetric, or constant-time varying) and the fact that dependence could vary according to the phase of the economic cycle to properly evaluate factors affecting the relationship.

The relationship between trip expenditure and length of stay has two important statistical regularities encountered in tourism research, which can justify a bivariate setting. On the one hand, there is a positive association between visitors' length of stay and total trip expenditure (Thrane, 2015). In this regard, the total expenditure per tourist is the product of the length of stay and daily expenditure. On the other, empirical findings have revealed an inverse association between daily expenditure and length of stay (e.g., García-Sánchez et al., 2013, Hellström, 2006 and Gómez-Déniz and Pérez-Rodríguez, 2021). In all cases, weak dependence is found.

In this chapter, we treat these two variables of interest separately and together. In the latter, the modeling will incorporate a bivariate distribution with discrete and continuous support. Additionally, two models will be shown, similar to those studied in the theory of collective risk, which allows both variables to be analyzed simultaneously, modeling aggregate tourist spending.

As usual, as it is of interest to economic agents related to tourism, we will implement a set of covariates that attempt to explain the behavior of tourists about these variables taken as a dependent variable.

3.2 Data

Our empirical analysis in this chapter focuses on Britons tourist visits during the year 2011 to any of the Canary Islands. 9797 observations have been chosen from the more than 100000 included in the original database, obtained from the Canary Islands Tourist Expenditure Survey (Encuesta de Gasto Turístico), which is based on personal interviews with tourists on departure and is carried out by the Canary Island Institute of Statistics (ISTAC) to obtain quarterly information about total tourist spending in the Islands. The original survey population is composed of domestic and foreign tourists who enter the Canary Islands at any of the airports in the archipelago. For the present study, those tourists whose expenditure in the country of origin

3.2. DATA

is zero were excluded. The survey compiles information from many nationalities, although most of the tourists in the Canary Islands are from European countries. The variable country of residence includes visitors from mainland Spain and Germany, Austria, Belgium, Denmark, Finland, France, Netherlands, Ireland, Italy, Norway, Poland, Portugal, United Kingdom, Czech Republic, Russia, Sweden, Switzerland, Luxembourg, and other countries (a general category including visitors from the USA and elsewhere). Among these 19 nationalities, 11 correspond to countries that form part of the Economic and Monetary Union (EMU), while eight are from non-euro countries. We obtained exchange rates against the euro for our analysis (see below). Here, we take into account both package and non-package tourists who stay for at least one night and no more than 30 consecutive nights. An example of the first twenty-five observations is shown in Table 3.1.

Table 3.2 shows descriptive statistics for the variables associated with the filtered database. These results show that the average tourist spent around 1444 euros in the country of origin (EO) and about 700 euro at the destination (ED). Furthermore, the mean of the daily expenditure at destination (DED) is about 85 euro. On the other hand, the expectation of their age is around 43 years old and that the family size was three persons. This average tourist had visited the Canary Islands three times (the number of times the respondent has previously visited the Canary Islands. A value of 0 is possible, indicating that at the moment of the interview, this is the tourist's first visit to the Canary Islands), had a household income (this variable is measured in the survey as an ordered categorical variable and not as a continuous variable. It takes the following values: 1, from €12000 to 24000; 2, from €24001 to 36000; 3, from €36001 to 48000; 4, from €48001 to 60000; 5, from €60001 to 72000; 6, from €72001 to 84000; and 7, higher than €84000) between 36000 and 48000 euro. About 50% of these tourists had arrived flying low cost (a dummy variable that takes the value 1 if the tourist visit was made using a low-cost carrier, and 0 otherwise), and 37% stayed at hotels of 4 or 5 stars (Aloj cat 1 is a dummy variable that takes the value 1 if the tourist accommodation is a hotel/aparthotel of 4-5 stars, and the value 0 otherwise. On the other hand, Aloj cat 2 is a dummy variable which takes the value 1 for a hotel/aparthotel of 1,2 or 3 stars, and 0 otherwise. More of the tourist belongs to class 4 followed by classes 3 and 8; regarding the job occupation (10 = Company owner, 9 = Self-employed/liberal profession, 8 = Skilled employee, 7 = Middle-level employee, 6 = Unskilled employee, 5 = Other worker, 4 = Student, 3 = Retired, 2 = Homemaker, 1 = Unemployed).

Table 3.2 also includes the bias and kurtosis values for the three expenditure variables. It is known (see, for instance, Groeneveld and Meeden, 1984) that the skewness's value using the standardized third central moment may become arbitrarily large and thus difficult to interpret. For this reason, we have also introduced the Bowley coefficient of skewness, $b_3 = (Q_3 + Q_1 - 2Q_2)/(Q_3 - Q_1)$, being Q_j the jth quartile of the data. Recall that $-1 < b_3 < 1$, with 1 representing extreme right skewness and –1, extreme left skewness. The moderate positive value of b_3 and the large kurtosis value for all variables in both cases suggest an empirical distribution moderated skewed to the right and with a long right tail. Thus, in all the cases,

Table 3.1: Example of the first twenty five observations.

alojcat12	alojcat3	age	EO	ED	income	lowcost	LS	job	persons	repetition
0	0	59	731.97	557.9	7	1	7	3	2	7
0	0	22	87.44	89.35	2	0	1	2	2	2
1	0	37	260.93	82.21	4	1	1	3	1	0
1	0	31	256.02	102.77	5	1	1	1	1	1
1	0	23	196.22	200.08	2	1	1	1	1	0
1	0	63	392.44	400.17	2	1	1	8	2	2
1	0	39	280.01	104.48	7	1	2	1	1	3
0	0	23	699	207.24	3	0	2	3	2	0
1	0	65	319.54	82.97	2	1	2	1	1	10
0	0	41	852.1	220.28	3	1	2	2	4	5
0	0	24	185.57	83.39	3	1	2	6	2	1
1	0	57	574.85	104.24	4	1	2	4	1	1
1	0	59	486.47	187.63	4	1	2	4	1	10
0	0	55	1147.11	364.82	1	1	2	9	3	0
0	1	28	226.84	51.38	1	1	2	7	1	0
1	0	48	3247.35	1436.09	5	1	2	9	8	8
0	1	30	284.29	202.9	1	1	2	6	1	0
0	0	40	700	504.57	7	1	2	3	2	3
1	0	21	182.99	104.47	1	1	3	7	1	2
1	0	49	487.94	94.04	6	1	3	6	2	9
0	0	50	1951.77	261.23	3	1	3	4	2	1
0	1	21	382.12	365.7	2	0	3	2	1	6
1	0	44	731.88	313.45	6	1	3	4	2	6
0	1	24	426.96	249.16	2	0	3	4	1	0
0	1	30	426.96	128.01	4	0	3	4	2	0

3.2. DATA

Table 3.2: Tourism data. Summary statistics for each variable. Filtered database.

Variables	Mean	St. Dev.	Min	Max	Relative frequency (dichotomous variables = 1)	Skewness	b_3	Kurtosis
alojcat12	0.369807	0.482777	0	1	36.98%			
alojcat3	0.129121	0.335351	0	1	12.91%			
age	42.9093	13.4485	16.	85.		2.593	0.201	14.99
Expenditure at origin (euro)	1443.32	1098.98	50.	12900.90		3.226	0.179	23.01
Expenditure at destination (euro)	704.48	600.724	2.01	7524.35		2.459	0.258	13.83
Total expenditure (euro)	2147.8	1392.9	176.79	15862.4				
Daily expenditure at destination (euro)	85.33	69.96	0.29	1838.95		4.863	0.097	68.16
income	3.41829	1.8901	1	7				
lowcost	0.481678	0.49969	0	1	48.17%			
LS	8.4777	3.24365	1	30				
job	4.143	2.35608	1	10				
persons	2.59967	1.19231	1	10				
repetition	3.33623	2.89493	0	10				

the expenditure is more concentrated below than above the mean but can consider enough large expenditure. These properties should be considered in the empirical modeling.

3.3 The length of stay variable

The length of stay (LS) is a key variable in a visitor's decision making process (Bull, 1995 and Decrop and Snelders, 2004; among others). Management of this question is of paramount importance for policymakers in the tourism industry since longer lengths of stay produce higher rates of hotel occupancy and hotel earnings. In addition, this question is closely related to the utility perceived by tourists during their holiday (Alegre et al., 2011). Studies in this field have highlighted the relationship between the length of stay and the income generated by tourists at their destination (Spotts and Mahoney, 1991; Taylor et al., 1993; Nogawa et al., 1996; Seaton and Palmer, 1997; Limburg, 1997; Mules, 1998; Agarwal and Yochum, 1999; Cannon and Ford, 2002; Alegre and Pou, 2003, Alegre et al., 2006; Martínez-García and Raya, 2008; Alegre et al., 2011; among others). Specifically, Martínez-García and Raya (2008) reported the influence of the price per day's stay and the tourist's level of income on the length of stay, as well that of the tourist's socio-demographic characteristics and of the type of stay.

Various econometric methods have been used to evaluate the determinants of length of stay, but most of them focus on an empirical statistical property such as the presence of two or more modes in the distribution of the length of stay. This property arises from tourists' propensity to take a holiday in weekly blocks (one, two, or three weeks are standard options in the sun and beach holidays).

Bimodal and multimodal distributions frequently arise in practice, particularly in cases involving a composite response from multiple, distinct sources and in aggregate survey data or user ratings when respondents give mixed opinions. Such distributions also arise in censored count data in which the highest category might create an additional mode (see Bose et al., 2013). The notion of multimodality of the distribution of a population is closely related to the notion of population heterogeneity. A popular way to model population heterogeneity parametrically is via mixture models. Since most classical discrete distributions can only capture unimodal empirical data, the discrete mixture of distributions is sometimes used to model this kind of empirical data. The disadvantage of this approach is that it can be complicated to introduce covariates into the resulting model. Various techniques have been proposed to address this issue. Thus, Nicolau and Más (2004), Alegre et al. (2006) and Alegre and Pou (2007) proposed estimating binary or multinomial logit models for intervals of less than seven days, of 7 to 14 days, and so on. However, a drawback of this type of modeling is that segmentation into weekly categories is rather arbitrary (Alegre et al., 2011). To overcome this, Alegre et al. (2011) proposed estimating a latent class Poisson regression model for the length of stay. This model assigns individuals endogenously to groups or classes with homogenous preferences. Each latent class corresponds to a sample segment that differs from the others in its

3.3. THE LENGTH OF STAY VARIABLE

preferences. For each class, the model can also estimate the specific effects of other variables that determine the final length of stay. Finally, Salmasi et al. (2012) used quantile regression to account for the multimodality of the length of stay. These authors estimated the price and income elasticities of the length of stay at different destinations in Italy using micro-level data.

Empirically, it can be shown that the length of tourist stay at some destinations (such as the Balearic Islands or the Canary Islands) not only presents bimodality but also overdispersion (the variance is larger than the mean). None of the classical distributions, such as Poisson, negative binomial, or generalized Poisson is capable of incorporating both features into the model. In addition, however, as tourist behavior is heterogeneous and model testing should consider this characteristic, we need to include both observed and unobserved heterogeneity in the model. In general, observed heterogeneity is easier to control, for example, when there are explicit assumptions about the variables and when spurious correlations can be avoided, but it becomes quite troublesome if the hypothesized cause-effect relationships are subject to variation over an unknown structure of sub-groups, or if they are simply unknown. In such cases, unobserved heterogeneity plays an important role because parameter estimates will incorporate bias and inconsistency. In this situation, we know from linear models that omitting a variable from a model introduces a bias into the OLS estimator, unless the omitted variable is uncorrelated with the other predictors in the model.

Bimodal and multimodal distributions are found in many continuous and discrete data sets. For example, in aggregate counts of responses to Likert scale questions, in online ratings of movies or hotels (Sur et al., 2015), or the duration of intervals between the eruptions of certain geysers. Also, in the distribution of male and female body weights, in student test scores, distinguishing between those who studied for the test and those who did not, or in tourism analysis, regarding the number of nights that tourists spend at a given destination (Alegre et al., 2011, Gómez-Déniz and Pérez-Rodríguez, 2019). However, these distributions have received little attention in the theoretical and empirical literature, except for classical distributions based on continuous data, such as exponential or normal distributions (Roeder, 1994), or discrete data frameworks such as censored count data in which an additional mode might be used for the highest category(see Bose et al., 2013), latent class models for count data which account for heterogeneity using a finite mixture of unimodal Poisson distributions (i.e., the latent class truncated Poisson regression; Wedel et al., 1993), or flexible models that capture both over and under-dispersion, such as the mixed Conway-Maxwell-Poisson distribution, which can reflect a wide range of truncated discrete data, and can exhibit either unimodal or bimodal behavior (Sur et al., 2015).

An essential feature of multimodal data sets is that they can reveal when two or more types of individuals are represented in a data set (for example, consumer segments and preferences). The discrete data observed in the specific case of the length of tourist stay at a significant "sun and sand" destination present not only bimodality but also overdispersion (the variance is greater than the mean). The length of stay is considered to be multimodal because tourists usually structure their trips in weekly blocks (for holidays in the Canary Islands, periods of one, two, or three weeks are the most common options). However, there is also some heterogeneity of tourist preferences regarding shorter or longer stays, which may depend on socioeconomic and demographic characteristics, the time available to tourists, and their prior familiarity with the destination, among other questions. Reflecting these diverse possibilities, empirical studies in this field have used several types of count data models. For example, latent class truncated Poisson models, to define different segments of tourists' preferences (Alegre et al., 2011) or count data quantile regression models to analyze the quantiles of the distribution of overall length of tourist stay (Salmasi et al., 2012), based on the Poisson distribution. More recently, Gómez-Déniz and Pérez-Rodríguez (2019) proposed two new statistical distributions to incorporate bimodality explicitly, including a flexible discrete distribution and a mixture model based on the Poisson distribution.

Figure 3.1 shows the histogram corresponding to the empirical length of stay variable. The pattern of bimodality can be seen at 7 and 14 days, which shows this data set.

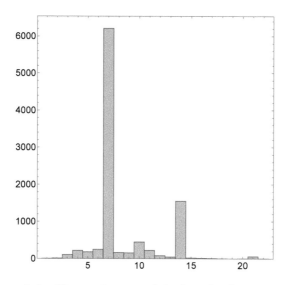

Figure 3.1: Observed count of the length of stay variable.

The total data are shown in Table 3.3.

3.3.1 Models

According to the standard literature in this field, briefly presented above, the length of tourist stay, T, like most expressions of the frequency of occurrence of an event, can be described by a Poisson distribution with mean $\lambda > 0$, i.e., the standard distribution for modeling random counts. However, in many cases, a model based on the Poisson distribution will be inadequate. Thus, in some situations, counts may occur

3.3. THE LENGTH OF STAY VARIABLE

Table 3.3: Observed counts for the variable length of stay.

LS	Frequency	LS	Frequency
1	5	16	11
2	12	17	10
3	107	18	4
4	221	19	1
5	182	20	5
6	250	21	59
7	6210	22	3
8	163	24	1
9	155	25	1
10	447	26	1
11	224	27	1
12	80	28	25
13	46	29	2
14	1546	30	2
15	23		
Total			9797

in clusters, giving rise to heterogeneity among individuals and provoking contagion (i.e., a degree of association between discrete events). When this happens, the count data may become overdispersed (i.e., the variance is greater than the mean), making the Poisson assumption very restrictive. On the other hand, if the parameter λ fits a gamma distribution with shape $r > 0$ and scale $(1-p)/p$, $0 < p < 1$, the unconditional distribution of T produces a negative binomial distribution with parameters r (dispersion parameter) and $1-p$. Nevertheless, although this distribution overcomes the problem of overdispersion, it still fails to reflect the bimodality observed in empirical data such as that for the length of tourist stay (usually expressed in days). The following result provides a reasonably well-suited distribution for modeling unimodal and bimodal data. Thus, it is appropriate for modeling the LS variable.

Theorem 3.1 *Let $g_Y(y; \mu, \sigma)$ be a discrete (or continuous distribution) with finite mean μ and variance σ^2. Then, it is verified that*

$$f_Y(y) = \omega(y; \mu, \sigma, \theta) g_Y(y; \mu, \sigma) \qquad (3.1)$$

for $-\infty < \theta < \infty$, where

$$\omega(y; \mu, \sigma, \theta) = \frac{1}{2 + \theta^2} \left[1 + \left(1 - \frac{\theta(y - \mu)}{\sigma}\right)^2 \right],$$

is a genuine probability mass function (density function in the continuous case).

Parameter θ controls the unimodality or bimodality of the family given in (3.1). Here $f_Y(y; \mu, \sigma)$ is the parent distribution from which we can construct a distribution that can be unimodal or bimodal. From the construction established in the previous result, it is apparent that this same result can be applied to obtain generalizations of classical distributions. The first candidates for this application, which rely on just a single parameter, would be the exponential distribution, for the continuous case, and the geometric and Poisson distributions, for the discrete case. In this paper, we consider the latter case. In other words, our starting point is that of a shifted Poisson distribution with parameter $\lambda > 0$. This situation is illustrated in the following result.

Proposition 3.1 *The expression given by*

$$f_T(t) = \omega_{\lambda,\theta}(t) \frac{\lambda^{t-1}}{(t-1)!} \exp(-\lambda) \qquad (3.2)$$

where $\lambda > 0$, $-\infty < \theta < \infty$ and

$$\omega_{\lambda,\theta}(t) = \frac{2\lambda + \theta(1+\lambda-t)\left[2\sqrt{\lambda} + \theta(1+\lambda-t)\right]}{\lambda(2+\theta^2)}$$

is a genuine probability mass function for $t = 1, 2, \ldots$

Proof: The proposition is an immediate consequence of applying the result provided in Theorem 3.1 to the shifted Poisson distribution with probability mass function (pmf) given by

$$g_T(t;\lambda) = \frac{\lambda^{t-1}}{(t-1)!} \exp(-\lambda), \quad t = 1, 2, \ldots$$

Hence the result. ∎

In order to achieve a more elegant expression for the above pmf, it is convenient to take $\lambda = \alpha^2$ and $\theta(1+\alpha^2-t) = \gamma_{\alpha,\theta}(t)$. The expression given in (3.2) can then be rewritten as

$$f_T(t) = \omega_{\alpha,\theta}(t) \frac{\alpha^{2(t-1)} \exp(-\alpha^2)}{(t-1)!}, \quad t = 1, 2, \ldots \qquad (3.3)$$

where $\omega_{\alpha,\theta}(t)$

$$\omega_{\alpha,\theta}(t) = \kappa(\theta) \left[2 + \frac{\gamma_{\alpha,\theta}(t)}{\alpha^2}(2\alpha + \gamma_{\alpha,\theta}(t))\right],$$

with $\kappa(\theta) = (2+\theta^2)^{-1}$. Figure 3.2 shows that the proposed distribution properly represents the unimodal or bimodal nature of empirical data. In this situation, the shifted Poisson distribution is a particular case for $\theta = 0$. Furthermore, it seems that as α tends to infinity and $\theta \to 0$, the normal distribution is an excellent approximation of the pmf (3.3). Some tedious but simple computations then provide the pgf of the distribution from which we get the moments of the distribution. In particular, the mean and the variance are given by

$$\begin{aligned}
E(T) &= 2\left[1 - \kappa(\theta)(1+\alpha\theta)\right] + \alpha^2, \\
var(T) &= \kappa(\theta)^2 \left[2(\theta^2 - \alpha\theta(2-\theta^2)) + \alpha^2(4(1+\theta^2) + 3\theta^4)\right],
\end{aligned}$$

3.3. THE LENGTH OF STAY VARIABLE

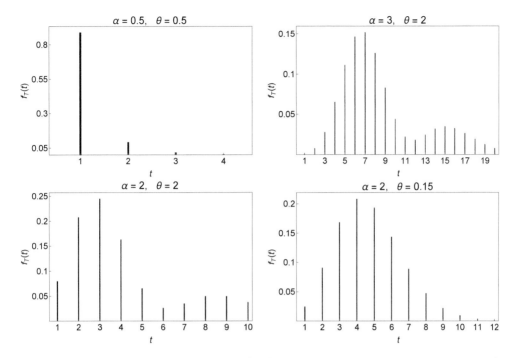

Figure 3.2: Graphs of the pmf given in (3.3) for special cases of parameters α and θ.

respectively. By determining the index of dispersion (ID) of a probability distribution, we can quantify the extent to which a set of occurrences is dispersed, compared to a standard pattern such as the Poisson distribution. The ID is the variance of a distribution divided by its mean. If the ID is greater than one, the corresponding distribution is said to be over-dispersed, and if it is less than one, the distribution is under-dispersed. In this case, the ID can take values greater or less than 1, so the distribution is appropriate to fit empirical data that present over- or under-dispersion.

Model estimation

Consider a sample with n observations $\tilde{t} = (t_1, t_2, \ldots, t_n)$, taken from the pmf (3.2). As a first approximation, the parameters α and θ can be estimated by the method of moments, assuming $\widehat{\mu} = \bar{t}$, where $\bar{t} = (1/n)\sum_{i=1}^{n} t_i$ is the sample mean. The estimation is then obtained by the maximum likelihood method. To do so, we first consider the model without covariates. Here, the log-likelihood function is proportional to

$$\ell(\tilde{t}; \alpha, \theta) \propto \sum_{i=1}^{n} \log \omega_{\alpha,\theta}(t_i) + 2n(\bar{t} - 1) - n\alpha^2. \tag{3.4}$$

The normal equations for estimating the parameters θ and α are given by

$$-\theta\kappa(\theta) + \frac{1}{n}\sum_{i=1}^{n} \frac{(1+\alpha^2 - t_i)(\alpha + \gamma_{\alpha,\theta}(t_i))}{2\alpha^2 + \gamma_{\alpha,\theta}(t_i)(2\alpha + \gamma_{\alpha,\theta}(t_i))} = 0, \tag{3.5}$$

$$-\alpha + \frac{\bar{t}-2}{\alpha} + \frac{1}{n}\sum_{i=1}^{n} \frac{2\alpha\theta(\alpha + \gamma_{\alpha,\theta}(t_i)) + 2\alpha + \gamma_{\alpha,\theta}(t_i)}{2\alpha^2 + \gamma_{\alpha,\theta}(t_i)(2\alpha + \gamma_{\alpha,\theta}(t_i))} = 0. \tag{3.6}$$

Equations (3.5)–(3.6) can be solved numerically for θ and μ using the Newton-Raphson iteration, for example starting from the seed point θ near to zero, with $\mu = \bar{t} + 1$.

Including covariates

Let us now assume that covariates are to be included in the model. First, consider that

$$\alpha(\theta, \mu) = \theta\kappa(\theta) + \sqrt{\mu - \varphi(\theta)\kappa(\theta)^2}, \tag{3.7}$$

where $\varphi(\theta) = 4 + \theta^2(5 + 2\theta^2)$ and where $\mu > \varphi(\theta)\kappa(\theta)^2$. Under this assumption, the mean of the pmf given in (3.3) is just the parameter μ, as is usually assumed when covariates must be included in the model. Now, let $\boldsymbol{x_i'} = [x_{1i}, x_{2i}, \ldots, x_{ki}]$ be a vector of $k \times 1$ covariates or factors associated with the length of stay of the ith tourist and where x_{ji} is the jth factor for the ith observation, $j = 1, 2, \ldots, k$. This vector of linearly independent regressors will determine t_i. The model provides great simplicity and the mean is straightforwardly expressed in terms of μ. Therefore, in order to introduce the covariates, we need only assume a translated one unit log link, defined by

$$\mu(\boldsymbol{x_i}, \boldsymbol{\beta}) = 1 + \exp(\boldsymbol{x_i'\beta}),$$

where $\boldsymbol{\beta} = (\beta_1, \ldots, \beta_k)'$ denotes the corresponding vector of regression coefficients. Furthermore, this logit link ensures that $\mu_i = \mu(\boldsymbol{x_i}, \boldsymbol{\beta})$ lies within the interval $[1, \infty)$. The log-likelihood of the model with covariates is similar to that given in (3.4) except that α is replaced by $\alpha(\theta, \mu)$, given in (3.7). Thus we have

$$\ell(\tilde{t}; \theta, \boldsymbol{\beta}) \propto \sum_{i=1}^{n} \log \omega_{\alpha(\theta,\mu_i)}(t_i) + 2\sum_{i=1}^{n}(t_i - 1)\log \alpha(\theta, \mu_i) - \sum_{i=1}^{n}[\alpha(\theta, \mu_i)]^2.$$

The marginal effect can be defined as the variation in the conditional mean of T caused by a one-unit change in the jth covariate. It is calculated as

$$\frac{\partial \mu_i}{\partial x_j} = \beta_j(\mu_i - 1),$$

for $i = 1, \ldots, n$ and $j = 1, \ldots, k$. The marginal effect indicates that a one-unit change in the jth regressor will increase or decrease the expectation of the length

3.3. THE LENGTH OF STAY VARIABLE

of stay. The effect is determined by the sign, positive or negative, of the regressor for each mean. For indicator variables such as x_k, which only take the value 0 or 1, the marginal effect in terms of the odds ratio is approximately $\exp(\beta_j)$. Therefore, when the indicator variable is one, the conditional mean is approximately $\exp(\beta_j)$ times greater than when the indicator is zero.

3.3.2 Numerical illustration

The proposed model was evaluated by the maximum likelihood method, without and also incorporating the survey information obtained for all tourists. Table 3.4 shows the results obtained for the model without covariates. The corresponding standard errors appear in brackets. This Table also includes the Akaike Information Criterion (AIC) and the number of observations. Comparisons were made with the shifted (one unity) Poisson (SP) and shifted (one unity) negative binomial (SNB) distributions. The results shown in 3.4 indicate that all the parameters are statistically significant at the 5% significance level and that model (3.2) is preferred to SP and SNB according to Vuong's closeness test. In this regard, we tested the null hypothesis that the two models are equally close to the actual model, against the alternative that one of them is closer (see Vuong, 1989). In the expression of the test statistics given in (1.26), $f(\cdot)$ and $g(\cdot)$ represent (3.2) and the alternative distributions, respectively.

Table 3.4: Maximum likelihood estimates and standard error (SE) in parenthesis for the data obtained by using (3.2) without including covariates.

Parameter	SP	SNB	BD
$\widehat{\theta}$	7.477	0.303	1.835
	(0.028)	(0.017)	(0.034)
$\widehat{\lambda}$		8.477	8.860
		(0.031)	(0.029)
AIC	49214.600	48732.400	46724.100
Observations	9797	9797	9797
Vuong's test			
SP vs. BD: 20.77 [0.00]			
SNB vs. BD: 27.25 [0.00]			

Figure 3.3 shows a smooth kernel distribution based on empirical data and the fitted pmf. The pattern of empirical data is captured by the proposed distribution.

Table 3.5 shows the results obtained with covariates and also, the ML estimation. This Table includes the corresponding coefficients and t statistics. The coefficients for covariates in the non-linear models should be interpreted with caution because the estimated coefficients are not the marginal effects, and so the marginal effects are computed using the formula: $\widehat{\beta}_k \exp(X'\widehat{\beta})$, where k represents the kth covariate, X' is a vector of covariates included in the mean equation and $\widehat{\beta}$ is a vector of estimated parameters.

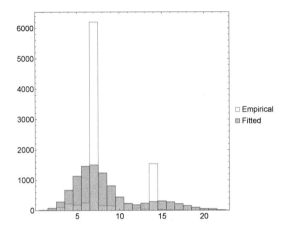

Figure 3.3: Observed and expected counts under the model with latent class without covariates.

Table 3.5: Results based on the bimodal distribution. The dependent variable is length of stay.

Variable	Estimate	t-Stat.
alojcat12	-0.129	13.10
alojcat3	-0.065	5.28
log(age)	-0.017	1.46
log(EO)	0.229	28.21
log(ED)	0.099	18.62
income	-0.021	9.44
lowcost	-0.036	4.46
job	0.013	7.39
persons	-0.058	14.06
repetition	0.015	11.11
constant	0.020	0.29
θ	0.682	29.65
AIC	45447.30	
Observations	9797	

Except for the constant and age, the rest of the variables are statistically significant. The expenditure at origin and destination, as was expected, has a positive effect on the stay. The larger the level of job and repetition (repetition of the trip was positively associated with length of stay), the longer the stay. The rest of the variables harm the stay. For example, more persons imply lower stay.

3.4 The expenditure variable

Most empirical studies of tourism demand use the micro-level (microdata) information and consider tourism expenditure per person per day as the dependent variable

(Wang and Davidson, 2010). The most widely used specifications of tourist spending models are those of dichotomous response, where the estimated coefficients indicate the probabilities that a tourist, in fact, makes the expenditure (Brida and Scuderi, 2012). Also, multinomial response models are used (Kim et al., 2010, Thrane, 2015 and Ferrer-Rosell et al., 2016), but they are not as usual as dichotomous responses. In general, the models that explain tourism demand and use spending as a dependent variable are supported by a set of covariates related to socioeconomic level, nationality, age, job, income, length of travel, type of travel, vacation accommodation, group travel, loyalty to the destination, among others. See Aguiló and Juaneda (2000), Fredman (2008), Craggs and Schofield (2009), Wang and Davidson (2010), Marcussen (2011), Thrane and Farstad (2011), Brida et al. (2013), García-Sánchez et al. (2013), Zheng and Zhang (2013), Thrane (2014), Marrocu et al. (2015), Disegna and Osti (2016), Aguiló et al. (2017) and Gómez-Déniz and Pérez-Rodríguez (2019). Recently, Gómez-Déniz et al. (2020) have used the beta regression model proposed in Ferrari and Cribari-Neto (2004) to explain the proportion of the expenditure at origin and destination.

The demand models that use spending are based on tourists' preferences and budget constraints (Brida et al., 2013). Lee and Choi (2019) explain, given that the tourist offer includes a series of goods and services that, due to their price, are exclusive for those tourists with sufficient resources and willing to make higher expenses, it is advisable to differentiate, for example, between luxury and regular items. Not to do so, it would produce a certain asymmetry in tourist spending. Different alternatives have been tried to solve the problem of asymmetry and obtain accurate estimates of tourism spending. The first, segmenting the market, Goryushkina et al. (2019) explains that a specifically identified group of clients with similar preferences and similar reactions is achieved by doing this. Craggs and Schofield (2009) segment tourism spending according to destination spending categories (shopping, coffee, bars) and identified statistically significant associations between different spending segments and a set of sociodemographic and behavioral variables. García-Sánchez et al. (2013) use different income segments and days of stay to estimate the determining factors of daily spending by foreign tourists. The second alternative is to disaggregate spending; Vinnciombe and Sou (2014) using tourism spending levels as a classification criterion. Alegre and Garau (2011) and Svensson et al. (2011) determine the tourist demand based on different tourist spending levels. Pani et al. (2020) study spending patterns in tourist establishments, distinguishing three spending levels: those who spend a lot, those who spend medium, and those who spend little. Gómez-Déniz et al. (2020) divide tourists' total expenditure per person and day according to their location, differentiating between expenses at the country of origin and expenses at destination. Lee and Choi (2019) examine the effect of different attributes on tourism spending in the destination. Attributes considered frustrating have a negative asymmetric effect, and good attributes have a positive asymmetric impact.

The third alternative is by studying the empirical distribution of data; Wu et al. (2013) use the scobit model that includes an asymmetry parameter that improves the logit and corrects possible skewness biases. When the estimated value of the asymmetry parameter is different from unity, the scobit model captures tourist spending distribution's asymmetry. Cárdenas et al. (2015) study spending as a function of tourists' degree of satisfaction; they assume that the variable spending per tourist and day follows a gamma distribution. Gómez-Déniz et al. (2020) studied tourist spending per person and day, related spending at destination and origin, and obtained a skewed distribution to the right that fits a beta-prime distribution.

The mentioned works agree that data on tourist spending per person and day have asymmetric behavior, and the estimates of average expenditure per person and day are likely biased. Little attention has been given to the asymmetry and the long right tail of the empirical data on which we focus in this paper. For that, Gómez-Déniz et al. (2021) proposed a reparameterization of the three-parameter log skew-normal distribution (LSN) for modeling the expenditure at the country of origin, destination, and total expenditure in the tourism setting. This distribution had been studied by Lin and Stoyanov (2009) (see also Azzalini, 2013, Chap. 2 and Azzalini et al., 2003) and recently applied in the actuarial setting by Gómez-Déniz and Calderín-Ojeda (2020). This proposal seems to fit the expenditure data satisfactorily in all the parts of the empirical distribution. In particular, the proposed model is well suited to capture the skewness and kurtosis that may be present and the long tail to the right that the three variables mentioned above tend to present in practice.

Graphs of the histogram of the expenditure data and the daily expenditure data (origin, destination, and both) are shown in Figure 3.4.

We have estimated the expenditure at density by using the pdf of the generalized gamma, inverse Gaussian and LSN distributions, given by,

$$
\begin{aligned}
f(x) &= \frac{\lambda \sigma^\mu}{\Gamma(\mu)} x^{\lambda\mu-1} \exp(-\sigma x^\lambda), \quad x>0,\ \mu>0,\ \sigma>0,\ \lambda>0, \\
f(x) &= \sqrt{\frac{\sigma}{2\pi x^3}} \exp\left[-\frac{\sigma(x-\mu)^2}{2x\mu^2}\right], \quad x>0,\ \mu>0,\ \sigma>0, \\
f(x) &= \frac{\phi(\eta_{\mu,\sigma}(x))}{\sigma x} \frac{\Phi((1+\eta_{\mu,\sigma}(x))\lambda)}{\Phi(\lambda_0)}, \quad x>0,
\end{aligned}
\tag{3.8}
$$

respectively, where

$$
\begin{aligned}
\eta_{\mu,\sigma}(x) &= \frac{\log x - \mu}{\sigma}, \\
\lambda_0 &= \frac{\lambda}{\sqrt{1+\lambda^2}}.
\end{aligned}
$$

The results are shown in Table 3.6 in which we have included some statistics to see which is the best model. They are the NLL, AIC, and CAIC. Furthermore, we also include the Kolmogorov-Smirnov test (KS), the Anderson Darling test (AD), and the Cramér-Von Mises test (CVM). Note that a model with lower statistics values is better than one with a higher value. All these results are shown in

3.4. THE EXPENDITURE VARIABLE

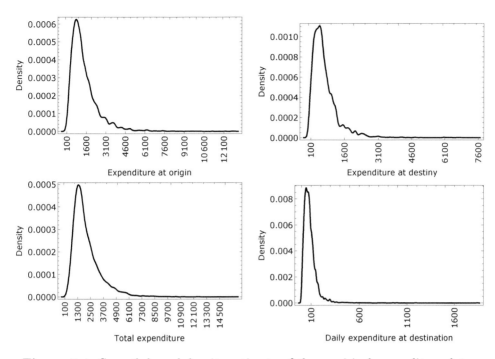

Figure 3.4: Smooth kernel density estimate of the empirical expenditure data.

Table 3.6. The corresponding p-values are very low, as usual, for the expenditure in all their versions. Nevertheless, the incorporated tests KS, CVM and AD, are lower for the LSN proposed.

Table 3.6: Parameters estimates, their p-values in brackets, maximum of the loglikelihood function, AIC and CAIC for the data expenditure at destination without including covariates

Distribution	$\widehat{\lambda}$	$\widehat{\mu}$	$\widehat{\sigma}$	NLL	AIC	CAIC	KS	CVM	AD
GG	0.612	4.400	0.084	73096.305	146199	146223	0.041	3.439	0.001
	[0, 00]	[0, 00]	[0, 00]						
IG		704.479	594.126	74535.429	149075	149091	0.136	58.00	0.014
		[0, 00]	[0, 00]						
LSN	-2.437	8.219	1.452	73062.058	146130	146155	0.028	2.002	5.82E-4
	[0, 00]	[0, 00]	[0, 00]						

Finally, the smooth kernel density estimate of the empirical expenditure data and the pdf of the LSN distribution obtained for estimated parameters provided in Table 3.6 is shown in Figure 3.5.

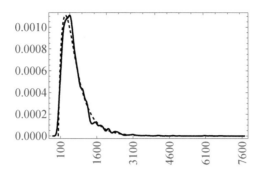

Figure 3.5: Smooth kernel density estimate of the empirical expenditure data and the pdf of the LSN distribution obtained for estimated parameters provided in Table 3.6.

3.5 Compound models

Our approach here is based on the relationship between aggregate expenditure and length of stay in a similar way to the relationship between the number of claims and the amount associated in the insurance scenario as studied in Chapter 2. Intuitively, expenditure X over period t is hypothesised to be the sum of $N(t)$ inter-period steps, Y_i, such that

$$X(N(t)) = \sum_{i=1}^{N(t)} Y_i.$$

Randomized time $N(t)$ can be thought of as the amount of economic time or the number of information events that agents actually experience during t, for example, the number of nights spent at a particular location.

If $\{X(t)\} \equiv \{X(t) : 0 < t < \infty\}$ is a stochastic process and $\{N(t)\}$, $N : \mathbb{N} \longrightarrow \mathbb{R}_+$ is a "driving process", then, the stochastically indexed $X(N(t))$ is "subordinated" to $X(t)$.

For example, consider that $\{N(t), t \geq 1\}$ is a discrete process (whether positive Poisson or positive negative binomial will be established later) for the length of stay at a vacation destination up to time t. In addition, let $\{Y_i\}_{i=1}^{\infty}$ be independent and identically distributed random variables representing individual expenditure, independent of $\{N(t), t \geq 1\}$.

To obtain a mathematically tractable model we have assumed the following: (i) that all the expenditures $Y_1, Y_2, \ldots, Y_{N(t)}$ are independent random variables with the same distribution function $F(y) = \Pr(Y \leq y)$; and (ii) that the random variable $N(t)$ is independent of all expenditures, $Y_1, Y_2, \ldots, Y_{N(t)}$ and follows a probability function $p_n = \Pr(N = n)$, $n = 1, 2, \ldots$. Although the hypothesis of independence between the variables $N(t)$ and Y_i may initially seem restrictive and incoherent, in fact the linear correlation between the random variables $X(t)$ and $N(t)$ is given by (see Sarabia and Guillén, 2008):

$$\rho(X(t), N(t)) = \left[1 + \frac{\text{C.V.}(X(t))}{\varrho(N(t))}\right]^{-1},$$

3.5. COMPOUND MODELS

where C.V.$(X(t))$ denotes the coefficient of variation of $X(t)$ and $\varrho(N(t))$ is the dispersion index of $N(t)$, assuming the existence of the corresponding moments.

Then, the cdf of the aggregate expenditure $X(t)$ is given by

$$F_{X(t)}(x) = \Pr(X(t) \leq x) = \Pr\left(\sum_{i=1}^{N(t)} Y_i \leq x\right). \quad (3.9)$$

In most cases, it is impossible to derive an explicit formula for (3.9), which is why it is necessary to assume independence between and within the two processes $\{N(t)\}$ and $\{Y_i\}_{i=1}^n$. In this case, we have the following fundamental relation,

$$F_{X(t)}(x) = \Pr(X(t) \leq x) = \sum_{k=1}^{\infty} p_k(t) F_Y^{*k}(x),$$

where F_Y^{*k} represents the kth convolution of F_Y. That is, $F_Y^{*k} = \Pr(Y_1 + Y_2 + \cdots + Y_k \leq x)$, the distribution function of the sum of k independent random variables with the same distribution as Y. In practice, this convolution can be computed by the expression

$$F_Y^{*k}(x) = \int_0^{\infty} F_Y^{*(k-1)}(x-y) F_Y(y) \, dy.$$

Furthermore, the pdf of the aggregate expenditure is given as

$$f_{X(t)}(x) = \sum_{k=1}^{\infty} p_k(t) f_Y^{*k}(x), \quad (3.10)$$

where $f_Y(x) = dF_Y(x)/dx$.

Determining the distribution of the total claims in an insurance portfolio in the context of risk theory is a question that has been addressed in many research studies. Today, this calculation is facilitated by modern computer technology, making it possible to process these convolutions relatively quickly and investigate the characteristic function using techniques such as the Fourier transform. Other methods such as the Escher transform, the Gram-Charlier approximation and the Edgeworth approximation are also greatly facilitated by computers. For more details on this, see Gerber (1979) and Rolski et al. (1999), among others.

Our aim in the present study is to obtain a closed-form expression for the aggregate expenditure to explain more than the tail probabilities, such as the factors that may affect total expenditure. For simplicity and because the convolution adopts a closed-form expression, we assume that Y_i ($i = 1, \ldots, n$) are n independent and identically distributed random variables that have exponential distributions with mean $1/\sigma$, $\sigma > 0$. Then, every Y_i has a pdf given by

$$f_Y(y) = \sigma \exp(-\sigma y) I_{(0,\infty)}(y), \quad y > 0,$$

where $I_{(0,\infty)}(y) = 1$ if $y > 0$ and 0 otherwise.

It is well-known that the sum of n independent exponential distributions with parameter $\gamma > 0$ produces a gamma variate with shape parameter n and rate parameter σ. That is,

$$f_Y^{*n}(y) = \frac{\sigma^n y^{n-1} \exp(-\sigma y)}{\Gamma(n)}, \quad y > 0. \tag{3.11}$$

Using conditional expectation arguments, it is straightforward to show that $\hat{\mathcal{L}}_{X(t)}(s) = \hat{g}_{N(t)}\left(\hat{\mathcal{L}}_Y(s)\right)$, where $\hat{g}_{N(t)}(\cdot)$ represents the pgf of the random variable $N(t)$ up to time t and $\hat{\mathcal{L}}_Y(s) = E[\exp(-sX(t))]$ is the Laplace-Stieljes transform of the random variable Y. By using (2.3) and (2.4) we get the mean and variance of $X(t)$, which are given by

$$E(X(t)) = E(N(t))E(Y), \tag{3.12}$$
$$var(X(t)) = E(N(t))var(Y) + var(N(t))E^2(Y). \tag{3.13}$$

Henceforth, we assume that the time horizon is fixed and so the time parameter is not necessary for the model.

3.5.1 The compound Poisson model

Empirically, the length of tourist stay usually presents non-zero observations. For that reason, assume a positive (truncated at zero) Poisson distribution for the randomised time N, such that

$$p_\mu(n) = p_n = \frac{\mu^n}{(\exp(\mu) - 1)n!}, \quad \mu > 0, \; n = 1, 2, \ldots \tag{3.14}$$

The mean and variance of this distribution are given by (3.30) and (3.31), respectively (see Problem 1 in this Chapter). It can be seen that the two curves are almost identical for $\theta \geq 4$.

Now, from (3.11) and applying (3.10) we obtain the density mixture of the convolution $X(N(t))$, such that

$$\begin{aligned} f_X(x) &= \sum_{k=1}^{\infty} \frac{\mu^k}{(\exp(\mu) - 1)k!} \frac{\sigma^k x^{k-1} \exp(-\sigma x)}{\Gamma(k)} \\ &= \sqrt{\frac{\sigma \mu}{x}} \frac{\exp[-(\sigma x + \mu)]}{1 - \exp(-\mu)} I_1\left(2\sqrt{\sigma \mu x}\right), \quad x > 0, \end{aligned} \tag{3.15}$$

where

$$I_\nu(z) = \sum_{k=0}^{\infty} \frac{(z/2)^{2k+\nu}}{\Gamma(k+1)\Gamma(\nu+k+1)}, \quad z \in \mathbb{R}, \; \nu \in \mathbb{R},$$

represents the modified Bessel function of the first kind (see, for example Abramowitz and Stegun, 1972, p. 360 for details).

Now, using (3.12), (3.13), (3.30) and (3.31), and taking into account that $E(Y) = 1/\gamma$ and $var(Y) = 1/\gamma^2$, the mean and the variance of the distribution given in (3.15) result,

$$E(X) = \frac{\mu}{\sigma(1 - \exp(-\mu))}, \qquad (3.16)$$

$$var(X) = \frac{\theta}{\sigma^2(1 - \exp(-\mu))}\left[1 + \frac{1 - (1 + \mu)\exp(-\mu)}{1 - \exp(-\mu)}\right].$$

3.5.2 The compound positive negative binomial model

The Poisson positive distribution, dependent on a single parameter, is a straightforward and somewhat inflexible model that may not adequately explain the length of tourist stay. It has been empirically proven that this variable presents the phenomenon of overdispersion (i.e., the variance is greater than the mean). For this reason, we consider a more flexible, two-parameter model that allows both over and under-dispersion (when the variance is less than the mean). This new model can arise in the following form. If $\mathcal{J}$ represents the length of stay in a tourism market (for example, the Canary Islands) and $\{\mathcal{I}_k; k = 1, \ldots, \mathcal{J}\}$ the lengths of stay of the tourists of different nationalities who come to the Canary Islands, and assume that $\mathcal{I}_k$, $k = 1, 2, \ldots$ are independent and identically distributed and do not depend on $\mathcal{J}$. Hence, $N = \sum_{k=1}^{\mathcal{J}} \mathcal{I}_k$, the total length of stay, is a compound distribution with a pgf that is given by $G_N(z) = G_{\mathcal{J}}[G_{\mathcal{I}}(z)]$, $|z| \leq 1$. It is easy to verify that the pgf of the truncated Poisson distribution in (3.14) gives $G_{\mathcal{J}}(z) = (\exp(z\theta) - 1)/(\exp(\theta) - 1)$. Now, if we assume that $\mathcal{I}_1$ follows a logarithmic series distribution (see Section 1.2.7 in this book) with parameter $0 < \iota < 1$ we have.

$$G_N(z) = G_{\mathcal{J}}[G_{\mathcal{I}}(z)] = \frac{1}{1 - \exp(\theta)}\left\{1 - \exp\left[\frac{\theta \log(1 - \iota z)}{\log(1 - \iota)}\right]\right\}$$

$$= \frac{1}{1 - \exp(\theta)}\left[1 - \exp(\theta)\left(\frac{1 - \iota}{1 - \iota z}\right)^{-\frac{\theta}{\log(1 - \iota)}}\right],$$

where we had taken into account that the pgf of the logarithmic series distribution is given by $G_{\mathcal{I}}(z) = \log(1 - \iota z)/\log(1 - \iota)$, $|z| \leq 1$.

If now we take $\mu = \iota$ and $\lambda = -\theta/\log(1 - \mu)$ we reach the following:

$$G_N(z) = \frac{\mu^\lambda}{1 - \mu^\lambda}\left[\frac{1}{(1 - \mu z)^\lambda} - 1\right],$$

which is the pgf of the truncated (at zero) negative binomial distribution with parameters $r > 0$ and $0 < p < 1$. To understand the role of the logarithmic distribution, we consider arguments similar to those made by Wani (1978), see also Johnson et al. (2005, Chap.7, p.303). Consider a group of N tourists containing M groups of different nationalities. If the expected frequency distribution of groups represented j

times is given by $E(f_j) = \xi \iota^j/j$, then $E(M) = -\xi \ln(1-\iota)$ and $E(N) = \xi\iota/(1-\iota)$. In consequence, the parameter ξ can be viewed as an index of diversity among tourist groups, i.e., a count of the number of different groups of tourists in the market considered.

Let us now assume a positive negative binomial distribution for N, such that

$$p_n = \frac{1}{1-\mu^\lambda}\binom{\lambda+n-1}{n}\mu^\lambda(1-\mu)^n, \quad \lambda > 0,\ 0 < \mu < 1,\ n = 1, 2, \ldots \quad (3.17)$$

The mean and the variance of this distribution are given by (3.32) and (3.33), respectively.

Again, we assume that Y_i ($i = 1, \ldots, t$) are n independent and identically distributed random variables that have exponential distributions with parameter $\sigma > 0$. After again using (3.11) and applying (3.10), we obtain the density mixture of the convolution $X(N(t))$, such that

$$\begin{aligned}f_X(x) &= \frac{\mu^\lambda}{1-\mu^\lambda}\frac{\exp(-\sigma x)}{x}\sum_{k=1}^{\infty}\frac{\Gamma(\lambda+k)}{\Gamma(k+1)\Gamma(\lambda)}\frac{(\sigma(1-\mu)x)^k}{\Gamma(k)}\\ &= \frac{\sigma\lambda(1-\mu)\mu^\lambda\exp(-\sigma x)}{1-\mu^\lambda}\,{}_1F_1(1+\lambda; 2; \sigma(1-\mu)x),\end{aligned} \quad (3.18)$$

for $x > 0$ and where ${}_1F_1(a; b; z)$ is the Kummer confluent hypergeometric function, given by

$$ {}_1F_1(a; b; z) = \sum_{j=0}^{\infty}\frac{(a)_j}{(b)_j}\frac{z^j}{j!},$$

being $(a)_j = \Gamma(a+j)/\Gamma(a)$ the Pochhammer symbol. See Abramowitz and Stegun (1972) and Gradshteyn and Ryzhik (1994) for details of this special function.

The mean and the variance of the distribution given in (3.18) are given by (3.34) and (3.35), respectively (see the problems proposed in this Chapter).

Numerical example

Table 3.7 shows the results obtained for the two models proposed, compound Poisson and compound negative binomial. In comparison with the results shown in Table 3.6 we conclude that the LSN model seems to be better. Nevertheless, when covariates are incorporated into the model (see Table 3.8), the compound negative binomial model outperforms the results provided with the LSN and compound Poisson models. To include covariates into the pdf (3.8) we have reparameterized this by writing μ as

$$\mu = -\frac{\sigma^2}{2} + \log\left[\frac{\zeta\,\Phi(\lambda_0)}{\Phi((1+\sigma)\lambda_0)}\right],$$

thus the resulting pdf given in (3.8) has now mean given by $\zeta > 0$.

3.5. COMPOUND MODELS

Table 3.7: Parameters estimates and *p*-values in brackets, maximum of the loglikelihood function, AIC and CAIC for the data expenditure at destination without including covariates.

Distribution	$\widehat{\lambda}$	$\widehat{\mu}$	$\widehat{\sigma}$	NLL	AIC	CAIC	KS	CVM	AD
CP		3.387 [0,00]	0.005 [0,00]	73491.885	146988	147004	0.064	10.504	0.0039
CNB	2.513 [0,00]	0.170 [0,00]	0.017 [0,00]	73329.770	146666	146690	0.078	14.587	0.0042

Table 3.8: Results based on LSN, compound Poisson and compound negative binomial models. Dependent variable, aggregate expenditure at destination.

	LSN		CP		CNB	
Variable	Estimate	*t*-Stat.	Estimate	*t*-Stat.	Estimate	*t*-Stat.
alojcat12	-0.452	-29.21	-0.226	-16.28	-0.243	-17.18
alojcat3	-0.347	-15.80	-0.173	-9.16	-0.191	-9.58
log(age)	0.007	0.32	-0.038	-2.09	-0.036	-1.87
income	0.024	6.71	0.033	9.97	0.033	9.28
lowcost	0.010	0.73	-0.015	-1.24	-0.017	-1.30
job	-0.012	-3.87	-0.014	-5.49	-0.013	-4.54
persons	0.196	35.74	0.216	41.96	0.213	36.52
repetition	0.017	6.93	0.019	9.27	0.019	8.10
μ			4.522	62.66	0.214	20.32
λ	0.006	10.76			3.482	33.56
σ	0.758	148.73				
constant	6.438	77.68	6.076	89.36	6.081	85.18
NLL	72475.021		72350.903		72246.598	
Observations	9797		9797		9797	

On the other hand, it was straightforward to include covariates in both models, CP and CNB, by equating (3.16) and (3.34) to γ, isolating σ and then applying the expression obtained in (3.15) and (3.18). The means of these two distributions are γ. Let $\boldsymbol{y_i'} = [y_{1i}, y_{2i}, \ldots, y_{ki}]$ be a vector of $k \times 1$ covariates or factors associated with the *i*th tourist's expenditure and let y_{ji} be the *j*th factor for the *i*th observation, $j = 1, 2, \ldots, k$. This is a vector of linearly independent regressors that we expect to determine X_i.

Then we used a log link defined by

$$\mu(\boldsymbol{y_i}, \beta) \equiv \gamma_i = \exp(\boldsymbol{y_i'}\beta),$$

where $\beta = (\beta_1, \ldots, \beta_k)'$, denotes the corresponding vector of regression coefficients. This log link ensured that γ falls in the interval $(0, \infty)$.

3.6 Bivariate model

We provide here a non-linear bivariate model for tourist expenditure and length of stay and daily tourist expenditure and length of stay which incorporates positive and negative dependence between these variables.

The total expenditure by an individual tourist i, $(i = 1, \ldots, I)$, E_i can be decomposed into personal daily expenditure (x_i) and length of stay (t_i). In other words,

$$E_i = x_i t_i.$$

As discussed above, both x_i and t_i have been modelled (separately) in the empirical literature. In this analysis, we seek a bivariate distribution for the random variables X and N (i.e., (X,T)) to model the dependence between them, assuming that their marginal distributions are gamma-type and censored (at zero) Poisson, respectively.

Empirically it can be shown that the correlation between daily tourist expenditure and the length of stay is negative. In contrast, the correlation between the expenditure (at origin, destination, or total) has a positive correlation with the variable length of stay.

Based on the Farlie-Gumbel-Morgenstern family of distributions (see Farlie, 1960 and Cambanis, 1977, among others) Gómez-Déniz and Pérez-Rodríguez (2021) propose to build a bivariate distribution with marginals given by

$$f_{\mu_1}(t) = \frac{1}{\mu_1}\left(1 - \frac{1}{\mu_1}\right)^{t-1}, \quad t = 1, 2, \ldots \tag{3.19}$$

$$f_{\mu_2,\beta}(x) = \frac{\beta^{\beta\mu_2}}{\Gamma(\beta\mu_2)} x^{\beta\mu_2 - 1} \exp(-\beta x), \quad x > 0. \tag{3.20}$$

They correspond to the shifted geometric distribution with parameter $\mu_1 > 1$ and a gamma distribution with the shape parameter $\beta\mu_2 > 0$ and rate parameter $\beta > 0$, respectively.

The cdf of the proposed distribution is given by

$$F_{\mu_1,\mu_2,\beta,\sigma}(t,x) = F_{\mu_1}(t) F_{\mu_2,\beta}(x) \left[1 + \sigma \bar{F}_{\mu_1}(t) \bar{F}_{\mu_2,\beta}(x)\right], \tag{3.21}$$

for $t = 1, 2, \ldots$, $x > 0$, where $\bar{F}_{\mu_1}(t) = (1 - 1/\mu_1)^{t-1}$ and $\bar{F}_{\mu_2,\beta}(x) = 1 - F_{\mu_2,\beta}(x)$ are the survival function of the random variables T and X, respectively.

3.6. BIVARIATE MODEL

Now, it is simple to see that

$$E_{\mu_1}(T) = \mu_1 > 1, \qquad (3.22)$$
$$var_{\mu_1}(T) = \sigma_T^2 = \mu_1(\mu_1 - 1), \qquad (3.23)$$
$$F_{\mu_1}(t) = \Pr(T \le t) = 1 - \left(1 - \frac{1}{\mu_1}\right)^t,$$
$$E_{\mu_2,\beta}(X) = \mu_2 > 0, \qquad (3.24)$$
$$var_{\mu_2,\beta}(X) = \sigma_X^2 = \frac{\mu_2}{\beta}, \qquad (3.25)$$
$$F_{\mu_2,\beta}(x) = 1 - \frac{\Gamma(\mu_2\beta, \beta x)}{\Gamma(\mu_2\beta)},$$

where $\Gamma(z)$ is the Euler gamma function defined in (1.5) and $\Gamma(a, z)$ is the incomplete gamma function defined in (1.10).

Let E_1 and E_2 be the set of all values of $F_{\mu_1}(n)$ and $F_{\mu_2,\beta}(x)$ with the exception of 0 and 1. By definition, a bivariate random variate (T, X) belongs to the FGM family of distributions if its joint cdf is given by (3.21), where (see Cambanis, 1977) $\sigma_{\inf} \le \sigma \le \sigma_{\max}$, being

$$\sigma_{\min} = -\min\left\{\frac{1}{M_1 M_2}, \frac{1}{(1-m_1)(1-m_2)}\right\},$$
$$\sigma_{\max} = \min\left\{\frac{1}{M_1(1-m_2)}, \frac{1}{(1-m_1)M_2}\right\},$$

with $m_1 = \inf E_1 = 1/\mu_1$, $m_2 = \inf E_2 = 0^+$, $M_1 = \sup E_1 = 1^-$ and $M_2 = \sup E_2 = 1^-$. Therefore, it follows that the acceptable value for the parameter σ is the interval $(-1, 1)$.

The pdf is obtained from (3.21). Its derivative with respect to x gives

$$F^*_{\Theta,\sigma}(t,x) = \frac{\partial F_{\Theta,\sigma}(t,x)}{\partial x}.$$

Second, the pdf is taken as

$$f_{\Theta,\sigma}(t,x) = F^*_{\Theta,\sigma}(t,x) - F^*_{\Theta,\sigma}(t-1,x),$$

from which, after simplifying, we obtain

$$f_{\Theta,\sigma}(t,x) = f_{\mu_1}(t) f_{\mu_2,\beta}(x) \left[f_{\mu_1}(t) H_{\Theta,\sigma}(x) + G_{\Theta,\sigma}(x)\right], \qquad (3.26)$$

where $\Theta = (\mu_1, \mu_2, \beta)$ and

$$H_{\Theta,\sigma}(x) = \sigma(1 - 2\mu_1)\left[1 - 2\bar{F}_{\mu_2,\beta}(x)\right],$$
$$G_{\Theta,\sigma}(x) = 1 + \sigma\left[1 - 2\bar{F}_{\mu_2,\beta}(x)\right].$$

The cross moment, $E_\Theta(TX)$, can be computed as

$$E_{\Theta,\sigma}(TX) = \int_0^\infty x \sum_{t=1}^\infty t f_\Theta(t,x)\,dx$$

$$= \int_0^\infty x f_{\mu_2,\beta}(x) \left[H_{\Theta,\sigma}(x) \sum_{t=1}^\infty t[f_{\mu_1}(t)]^2 + \mu_1 G_{\Theta,\sigma}(x) \right] dx.$$

Tedious but straightforward computations then provide

$$\sum_{t=1}^\infty t[f_{\mu_1}(t)]^2 = \left(\frac{\mu_1}{2\mu_1 - 1}\right)^2,$$

$$\int_0^\infty x f_{\mu_2,\beta}(x) \bar{F}_{\mu_2,\beta}(x)\,dx = M(\Theta),$$

in which we use the following relation between the incomplete gamma function and the Kummer confluent hypergeometric function,

$$\Gamma(a,z) = \Gamma(a) - \frac{z^a}{a}\,_1F_1(a; a+1; -z)$$

and the fact that

$$_2F_1(a,b;c;z) = \frac{1}{\Gamma(b)} \int_0^\infty t^{b-1} \exp(-t)\,_1F_1(a;c;tz)\,dt.$$

Finally, the cross moment is obtained as

$$E_\Theta(TX) = \mu_1\mu_2 \left[1 + \frac{\sigma(\mu_1 - 1)(2M(\Theta) - 1)}{1 - 2\mu_1} \right],$$

while the covariance is given by

$$\operatorname{cov}(T, X) = \frac{\mu_1\mu_2\sigma(\mu_1 - 1)(2M(\Theta) - 1)}{1 - 2\mu_1}.$$

The marginal distributions of X and T are given by (3.19) and (3.20), respectively, with marginal means and variances given by (3.22), (3.23), (3.24) and (3.25), respectively. Simple calculation then provides the cross moment, $E(TX)$, the covariance and the coefficient of correlation, which is given by

$$\varrho(T, X) = \frac{\sigma\beta\sigma_N\sigma_X\,[2M(\Theta) - 1]}{1 - 2\mu_1},$$

where

$$M(\Theta) = \frac{\Gamma(2\mu_2\beta + 1)}{\Gamma(\mu_2\beta)\Gamma(\mu_2\beta + 2)}\,_2F_1(\mu_2\beta + 1, 2\mu_2\beta + 1; \mu_2\beta + 2, -1).$$

3.6. BIVARIATE MODEL

Here $_2F_1$ is the hypergeometric function. This correlation can take positive, negative, or zero values, depending on the value and the sign of the parameter σ controlling the dependence (positive or negative) or independence ($\sigma = 0$) of the model.

The conditional distribution of X given $T = t$ and the conditional distribution of T given $X = x$ are given by

$$\begin{aligned}
f_{\Theta,\sigma}(t|X=x) &= f_{\mu_1}(t)\left[f_{\mu_1}(t)H_{\Theta,\sigma}(x) + G_{\Theta,\sigma}(x)\right], \\
f_{\Theta,\sigma}(x|T=t) &= f_{\mu_2,\beta}(x)\left[f_{\mu_1}(t)H_{\Theta,\sigma}(x) + G_{\Theta,\sigma}(x)\right],
\end{aligned}$$

and the conditional means are then

$$\begin{aligned}
E_{\Theta,\sigma}(T|X=x) &= \mu_1\left[1 + \frac{(\mu_1-1)\sigma}{2\mu_1-1}\left(1 - 2\bar{F}_{\mu_2,\beta}(x)\right)\right], \\
E_{\Theta,\sigma}(X|T=t) &= \mu_2\left[1 + \sigma M(\Theta)\left(1 - (2\mu_1 - 1)f_{\mu_1}(t)\right)\right].
\end{aligned}$$

Note the dependence of these conditional means on t and x, respectively. This dependence can be positive or negative according to the sign of the parameter σ. The versatility of the model enables it to explain the negative dependence empirically revealed between daily tourist expenditure and length of stay, as well as the positive dependence that empirically appears to exist between aggregate expenditure (at origin or destination or both simultaneously) and length of stay.

3.6.1 Some methods of estimation

Let us first consider the model with no covariates, in which

$$(\tilde{t}, \tilde{x}) = \{(t_1, x_1), \ldots, (t_I, x_I)\}$$

is a sample obtained from the distribution (3.26) and $\bar{t} = (1/I)\sum_{i=1}^{I} t_i$, $\bar{x} = (1/I)\sum_{i=1}^{I} x_i$, $s_X^2 = var(X)$ and $\varrho(T,X)$ are the corresponding sample moments, variances and correlation, respectively. The estimators based on these sample moments are then computed sequentially as follows.

$$\begin{aligned}
\widehat{\mu}_1 &= \bar{t}, \quad \widehat{\mu}_2 = \bar{x}, \quad \widehat{\beta} = \frac{\widehat{\mu}_2}{s_X^2}, \\
\widehat{\sigma} &= \frac{(1-2\widehat{\mu}_1)\varrho(T,X)}{\widehat{\beta}\sigma_T\sigma_X(2M(\widehat{\Theta})-1)}.
\end{aligned} \quad (3.27)$$

When the maximum likelihood method is used as the estimation method, we first consider the case in which the model has no covariates and we aim to estimate its four parameters, μ_1, μ_2, β and σ. Maximizing the log-likelihood function concerning these four parameters could be unsatisfactory because certain combinations of parameters might violate the restriction of the support of the σ parameter. To avoid this situation, we propose a two-step estimation method. Firstly, σ is estimated from

(3.27) by the method of moments. This estimated value, say $\widehat{\sigma}$, is then incorporated into the log-likelihood function to estimate the remaining model parameters, by the maximum likelihood method, i.e., by maximizing the log-likelihood function, which is proportional to

$$\ell((\tilde{t}; \tilde{x}); \Theta) \propto I\left[-\log \mu_1 + (\bar{t}-1)\log\left(1 - \frac{1}{\mu_1}\right)\right.$$
$$\left. + \beta\left(\mu_2\left(\log \beta + \sum_{i=1}^{I} \log x_i\right) - \bar{x}\right) - \log \Gamma(\mu_2 \beta)\right]$$
$$+ \sum_{i=1}^{I} \log\left[f_{\mu_1}(t_i) H_{\Theta, \sigma}(x_i) + G_{\Theta, \widehat{\sigma}}(x_i)\right]. \quad (3.28)$$

The normal equations obtained from (3.28) require the use of the digamma function, $\psi(z) = \frac{d}{dz}\log(\Gamma(z))$, $z > 0$, in order to estimate all the model parameters. However, this problem is overcome by means of Mathematica routines (see Ruskeepaa, 2009) and RATS (see Brooks, 2009), which work well with this special function. Other packages, such as MATLAB® and R, can also be useful. Moreover, the dependency of the gamma function can be overcome by replacing this function using the following approximation,

$$\log(\Gamma(z)) \approx \frac{1}{2}\log(2\pi) + \left(z - \frac{1}{2}\right)\log(z) - z + \frac{z}{2}\log\left(z \sinh\left(\frac{1}{z}\right)\right),$$

which is well-known in the statistical literature.

Let us consider a more realistic model in which covariates are included (e.g., trip-related characteristics or socio-demographic variables). When regression analysis is to be performed, it is often helpful to model the mean of the response.

Now, let $\boldsymbol{y}_i = (y_{1i}, \ldots, y_{ki})'$ and $\boldsymbol{z_i} = (z_{1i}, \ldots, z_{ki})'$ be two vectors of k covariates associated with the ith observation. These vectors of linearly independent regressors are assumed to determine (t, x). For the ith observation, the model takes the form:

$$(T_i, X_i) \sim f_{T,X}(t_i, x_i; \beta, \widehat{\sigma}, \mu_{1i}, \mu_{2i}),$$
$$\log(\mu_{1i} - 1) = \boldsymbol{z}_i'\boldsymbol{\eta},$$
$$\log(\mu_{2i}) = \boldsymbol{y}_i'\boldsymbol{\delta},$$

for $i = 1, \ldots, I$ and where I denotes the number of observations and $\boldsymbol{\delta} = (\delta_1, \ldots, \delta_k)'$ and $\boldsymbol{\eta} = (\eta_1, \ldots, \eta_k)'$ the corresponding vectors of regression coefficients. For convenience we have written $\mu_{1i} = \mu_1(\boldsymbol{z_i}, \boldsymbol{\eta})$ and $\mu_{2i} = \mu_2(\boldsymbol{y_i}, \boldsymbol{\delta})$. In principle, each of the variables, N and X, can be influenced by different factors, and so the explanatory variables that are taken to explain $\mu_{\kappa i}$, $\kappa = 1, 2$, are not the same. Furthermore, the logit link assumed ensures that μ_{1i} falls within the interval $(1, \infty)$ and μ_{2i} is within the interval $(0, \infty)$.

3.6. BIVARIATE MODEL

Now, let $\Theta = (\beta, \boldsymbol{\eta}, \boldsymbol{\delta})$. The log-likelihood function is then proportional to

$$\ell((\tilde{t}; \tilde{x}); \Theta) \propto -\sum_{i=1}^{I} \log \mu_{1i} + \sum_{i=1}^{I} (t_i - 1) \log \left(1 - \frac{1}{\mu_{1i}}\right)$$

$$-\sum_{i=1}^{I} \log \Gamma(\mu_{2i}\beta) + \beta \left[\log \beta \sum_{i=1}^{I} (\mu_{2i}(1 + \log x_i)) - \bar{x}\right]$$

$$+\sum_{i=1}^{I} \log \left[f_{\mu_{1i}}(t_i) H_{\Theta, \hat{\sigma}}(x_i) + G_{\Theta, \hat{\sigma}}(x_i)\right].$$

Numerical illustration

We have used the empirical correlation between the variables LS and ED and LS and DED which result 0.340 and -0.083, respectively, in order to estimate, by using equation (3.27), the values of the parameter σ. We got $\hat{\sigma} = 1.129$ and $\hat{\sigma} = -0.279$, respectively. They are within acceptable limits into the interval $(\sigma_{\min}, \sigma_{\max})$. Table 3.9 includes parameter estimates and standard errors (SE), the negative value of the maximum of the log-likelihood function (NLL), and the number of observations corresponding to the two variables modeled in the bivariate distribution: daily expenditure (at destiny) and length of stay and expenditure (at destiny) and length of stay. As we can see, the parameters results are all significant.

Table 3.9: Results based on the Farlie-Gumbel Morgenstern copula without including covariates.

	LS & ED		LS & DED	
Variable	Estimate	SE	Estimate	SE
$\hat{\mu}_1$	8.269	0.076	8.549	0.081
$\hat{\mu}_2$	6.138	0.01	4.212	0.010
$\hat{\beta}$	8.100	0.110	5.467	0.077
NLL	42319.951		42526.875	
Observations	9797		9791	

Figure 3.6 (left) shows the empirical smoothed bivariate distribution and the adjusted distribution (right), calculated according to the maximum likelihood estimators listed in Table 3.9. In general, and based on inspection of the graphs of the bivariate distribution, the adjusted distribution has the pattern of the bivariate distribution and perfectly represents the value of the mode. Table 3.10 summarises the results obtained for our initial bivariate model including covariates and simultaneity affecting the mean of each variable. The other dependent variable is included in the μ_1 and μ_2 equations as a way to include simultaneity in the mean equations.

Analysis of the results shown in Table 3.10 reveals similar results in terms of the variable ED in comparison with the ones provided in 3.8. Different effects are produced on LS and DED and LS and ED.

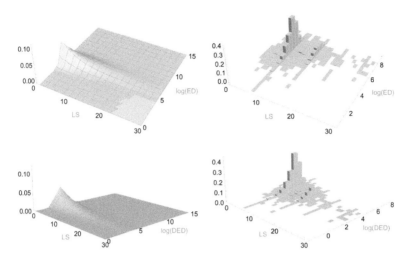

Figure 3.6: Empirical smooth distribution (left) and fitted model (right).

The following observations can be made concerning factors explaining the duration of tourist stay and the amount of holiday expenditure. Regarding the determinants of length of stay, the coefficient of daily expenditure is not statistically significant. The coefficients for the income variables are close to zero and statistically significant. For example, Hellström (2006) obtained a negative value, but it was not statistically significant. However, other analysts, such as Wang et al (2012), have reported a positive association in this respect, which is in line with the demand theory.

Regarding the influence of personal characteristics, repetition of the trip was positively associated with length of stay, although Thrane (2012) measured a negative effect for this parameter and a positive association with the ED but not with the DED.

Finally, variables such that log(age) are positively associated with duration and negatively with the DED.

High income has a more substantial effect on the ED and DED. In other words, tourists with higher incomes have a greater propensity to spend than those with lower incomes (reference category). These results are in line with the fact that higher household income has been positively associated with higher levels of spending (Aguiló et al., 2017) and that high-income tourists spend 50% more than low-income tourists (García-Sánchez et al., 2013). Similar results have also been found by Gómez-Déniz and Pérez-Rodríguez (2019). Nevertheless, the income is negatively associated with the LS.

Statistically-significant negative effects were found for the coefficients related to accommodation and job. In this last case, only for the expenditure variable. The signs obtained for the occupation coefficients are counterintuitive, suggesting that the status of the business owner or salaried worker is inversely associated with the ED and DED, concerning the reference category but is positively related to the LS.

3.6. BIVARIATE MODEL

Table 3.10: Results based on the Farlie-Gumbel Morgenstern copula.

Variable	LS Estimate	LS t-Stat.	ED Estimate	ED t-Stat.	LS Estimate	LS t-Stat.	DED Estimate	DED t-Stat.
alojcat12	-0.052	-2.34	-0.078	-24.10	-0.040	-1.77	-0.117	-28.55
alojcat3	-0.030	-0.96	-0.057	-13.36	-0.032	-0.90	-0.088	-13.88
log(age)	0.165	5.18	0.005	1.16	0.158	4.68	-0.030	-4.92
income	-0.012	-2.24	0.003	4.80	-0.010	-1.75	0.007	6.93
lowcost	-0.064	-3.29	0.002	0.66	-0.066	-3.18	0.017	4.26
job	0.012	2.85	-0.002	-3.33	0.014	3.15	-0.005	-6.57
persons	0.045	5.09	0.030	28.24	0.045	5.35	0.035	23.41
repetition	0.015	4.29	0.002	5.92	0.015	4.33	9.6E-4	1.32
β	9.621	10.39	66.82	107.12	6.437	9.96	66.59	64.44
constant	1.240		1.737		1.284		1.493	
NLL	41428.154				41667.785			
Observations	9797				9791			

3.7 Generalized additive model

The Ordinary Least Squares (OLS) method occupies a prominent place in the field of applied statistics. It is a simple model, easy to estimate, and with interpretations that are easy to implement in any scenario. With the explosive development of computational methods a few decades ago, an alternative to it was proposed, the generalized additive model, a generalization of the linear regression model, which consists of replacing the linear relationship of the covariates with a smoothed function, generally of the polynomial character. The sum of these functions constitutes an additive model (hence the name by which it is known). These models are known in the literature as generalized additive models (GAM). It is not an objective of this text to detail these models. For this, the reader can read the basic references in the matter that constitute Hastie and Tibshirani (1986), Yee (2015) and Wood (2017).

We start by estimating the basic OLS model and a GAM model to compare the results with the ones provided in Table 3.5. We have chosen to fit a basic GAM model, given by $Y_l = \beta_0 + \sum_j f_j(x_l) + u_l$, where $f_j(\cdot)$ is the smoothing spline for the independent variables x_l and u_l, independent normal random variates. Many statistical packages, especially in R, allow the incorporation of a parametric model for the response variable through the appropriate link (see Wood, 2017), whose work constitutes an irreplaceable reference in the matter. We have used the *mgcv* package with cubic regression splines. The results for the OLS and GAM models obtained are shown in Table 3.11. The GAM model includes three terms to be smoothed, age, EO, and ED. Observe that the rest of the variables are dichotomous or categorical.

Regarding the results obtained from the estimation of these two models, see Table 3.11, both provide the same effects on significance and sign for the explanatory variables in origin, destination, and total expenditure. The results for the smooth terms are summarized in this Table by the effective degrees of freedom (EDF), which measures the complexity of a penalized smooth term. As is well-known, EDF can be interpreted as an estimate of how many parameters are needed to represent the smooth. If the EDF equals 1, a linear relationship cannot be rejected. In our study, the EDFs estimated show clearly non-linearity (see also Figure 3.7). The best fit to data among these models correspond to the GAM model since it provides a lower value for the NLL and AIC.

For the age variable, it is observed in this Figure that approximately below 42 years, the number of days of stay is below the mean and above for age greater than 42 years. Regarding the expenditure at destination, a value of approximately 665.15 euros makes the difference between a stay below or above the average. This value increases to approximately 1096 euros for the variable expenditure at origin.

Besides being continuous variables, concerning age García-Sánchez et al. (2013) ensure that spending may decrease for the age among the oldest and the youngest tourists. There is a U-inverted-shaped relationship between tourist age and daily expenditure. Regarding the length of stay, average expenditure per person per day diminishes for longer trips, due to the economies of scale, in this line coincide Thrane and Farstad (2011), and Thrane (2014).

3.7. GENERALIZED ADDITIVE MODEL

Table 3.11: Results based on OLS and GAM models. Dependent variable, LS.

	OLS		GAM	
Variable	Estimate	p-value	Estimate	p-value
alojcat12	-0.120	< 0.001	-0.125	< 0.001
alojcat3	-0.047	< 0.001	-0.042	< 0.001
log(age)	0.080	< 0.001		
log(EO)	0.244	< 0.001		
log(ED)	0.110	< 0.001		
income	-0.017	< 0.001	-0.017	< 0.001
lowcost	-0.007	0.201	-0.009	0.112
job	0.012	< 0.001	0.010	< 0.001
persons	-0.063	< 0.001	-0.075	< 0.001
repetition	0.010	< 0.001	0.009	< 0.001
constant	-0.440	< 0.001	2.314	< 0.001
Approximate significance of smooth terms				
			EDF	p-value
s(log(age))			1.962	< 0.001
s(log(EO))			1.991	< 0.001
s(log(ED))			1.995	< 0.001
Observations	9797		9797	
NLL	1581.370		1396.649	
AIC	3186.741		2823.192	

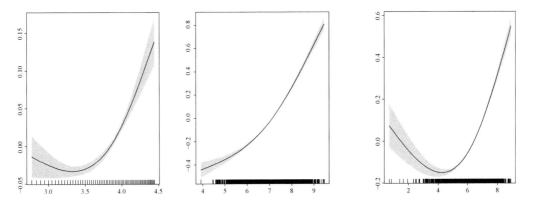

Figure 3.7: Fitted functions for the smoothed variables in the GAM model. From top to down and left to right we have log(EO), log(ED) and log(Age).

Exercises

1. Show that for the translated one unit log link, defined by

$$\mu(\boldsymbol{x_i}, \boldsymbol{\beta}) = 1 + \exp(\boldsymbol{x_i'}\boldsymbol{\beta}), \tag{3.29}$$

where $\boldsymbol{\beta} = (\beta_1, \ldots, \beta_k)'$ denotes the corresponding vector of regression coefficients, $\mu_i = \mu(\boldsymbol{x_i}, \boldsymbol{\beta})$ lies within the interval $[1, \infty)$.

2. Show that the marginal effect for the link (3.29) is given by
$$\frac{\partial \mu_i}{\partial x_j} = \beta_j(\mu_i - 1), \ i = 1, \ldots, n; \ j = 1, \ldots, k.$$

3. Show that for indicator variables such as x_k, which only take the value 0 or 1, the marginal effect in terms of the odds ratio is approximately $\exp(\beta_j)$. Therefore, when the indicator variable is one, the conditional mean is approximately $\exp(\beta_j)$ times greater than when the indicator is zero.

4. Show that the mean and variance of the distribution given in (3.14) are given by
$$E(N) = \frac{\mu}{1 - \exp(-\mu)}, \tag{3.30}$$
$$var(N) = \frac{\mu(1 - (1 + \mu)\exp(-\mu))}{(1 - \exp(-\mu))^2}, \tag{3.31}$$
respectively.

5. Show that the mean and variance of the truncated negative binomial distribution given in (3.17) are given by
$$E(N) = \frac{\lambda(1 - \mu)}{\mu(1 - \mu^\lambda)}, \tag{3.32}$$
$$var(N) = \frac{\lambda\mu[1 - \mu^\lambda(1 + (1 - \mu)\lambda)]}{\mu^2(1 - \mu^\lambda)^2}, \tag{3.33}$$
respectively

6. Using (3.32) and (3.33) to get that the mean and variance of the distribution provided in (3.18) are given by,
$$E(X) = \frac{\lambda(1 - \mu)}{\sigma\mu(1 - \mu^\lambda)}, \tag{3.34}$$
$$var(X) = \frac{\lambda(1 - \mu)}{\sigma^2\mu(1 - \mu^\lambda)}\left[1 + \frac{1 - \mu^\lambda(1 + (1 - \mu)\lambda)}{\mu(1 - (1 - \mu)^\lambda)}\right]. \tag{3.35}$$

7. Show that the pgf of the truncated Poisson distribution in (3.14) is given by $G(z) = (\exp(z\theta) - 1)/(\exp(\theta) - 1), |z| \leq 1$.

8. Show that the sum of n independent exponential distributions with parameter $\gamma > 0$ produces a gamma variate with shape parameter n and scale parameter γ, with pdf given by
$$f(y) = \frac{\gamma^n y^{n-1} \exp(-\gamma y)}{\Gamma(n)}, \ y > 0.$$

3.7. GENERALIZED ADDITIVE MODEL

9. Under the compound model, assuming the existence of the corresponding moments, show that the linear correlation between the random variables $X(t)$ and $N(t)$ is given by

$$\rho(X(t), N(t)) = \left[1 + \frac{\text{C.V.}(X(t))}{\varrho(N(t))}\right]^{-1},$$

where C.V.$(X(t))$ denotes the coefficient of variation of $X(t)$ and $\varrho(N(t))$ is the dispersion index of $N(t)$.

10. Show that (3.26) represents a genuine bivariate pdf if $\sigma_{\inf} \leq \sigma \leq \sigma_{\sup}$, where

$$\sigma_{\inf} = -\min\left\{\frac{1}{M_1 M_2}, \frac{1}{(1-m_1)(1-m_2)}\right\},$$

$$\sigma_{\sup} = \min\left\{\frac{1}{M_1(1-m_2)}, \frac{1}{(1-m_1)M_2}\right\},$$

being $m_1 = \inf E_1 = 1/\mu_1$, $m_2 = \inf E_2 = 0^+$, $M_1 = \sup E_1 = 1^-$ and $M_2 = \sup E_2 = 1^-$.

Chapter 4

Statistical Distributions in Other Fields

4.1 Introduction

This chapter briefly addresses four areas of economics that have attracted much interest in recent decades from the research community. These are stochastic frontier models, Geography, duration models, and income distribution models.

We first devote our attention to stochastic frontier models. These models are concerned with estimating technical efficiency for a specific assumed production function, generally the Cobb-Douglas production function. The estimation of the production or cost functions must respect the fact that actual production cannot exceed the maximum possible production given the quantities of inputs. In this sense, the work consists of relating the inefficiency with some factors that are likely to be determinants and measuring what measures contribute to inefficiency. Under some distributive assumptions, the idiosyncratic error can be expressed as a function of specific covariates, obtaining a closed-form likelihood function. The maximum likelihood method can get parameter estimates and provide inefficiency measures.

The distribution of the size of cities constitutes an essential element in Geography to try to explain the urban growth of cities. It is well known that in the case of city populations, the resulting distribution in a country, a region, or even the world is characterized by its largest city, with other cities decreasing in size, initially at a faster rate and then more slowly. There are a few large cities and a much large number of cities of smaller magnitude. In this sense, their size distribution constitutes a crucial element in the regional organization for their location and the location of features that citizens usually demand, and statistics also play a prominent role in this field.

Autoregressive conditional duration models are a fundamental tool for dealing with data that arrives at regular time intervals, as occurs with transactions in financial markets. The time between the different events is treated as a stochastic process with dependent arrival rates, distinguishing between the number of aggregated transactions, the total volume traded, and daily price movement.

Finally, in this chapter, we dedicate some pages of it to studying income distribution, which has a long history in economics and statistics. The star instrument to carry out this study is the Lorenz curve, which measures the accumulated proportion of income with the lowest percentage of the population. Its graphical representation offers a complete picture of the concentration in distribution, and its popularity has spread to other fields of economics, such as bibliometrics and physics. The fundamental role of these curves in analyzing inequality has led to the development of new functional forms to explain income. Thus, it has given rise to an essential body of research in this area.

4.2 Stochastic frontier analysis

The main objective of the stochastic frontier approach is to estimate the level of technical efficiency for some production function. Technical efficiency is defined as the ratio of the observed output to the corresponding frontier and which is estimated from the composed error term. In the usual stochastic frontier model, it is acknowledged that the estimation of production or cost functions must respect the fact that actual production cannot exceed the maximum possible production given input quantities. The task is to relate inefficiency to some factors that are likely to be determinants and measure how they contribute to inefficiency. Under some distributional assumptions, the idiosyncratic error can be expressed as a function of specific covariates, and a closed-form likelihood function can be derived. Finally, the method of maximum likelihood may be used to obtain parameter estimates and provide inefficiency measures.

The classical stochastic frontier model (see Aigner et al., 1977, Meeusen and Broeck, 1977, Battese and Coelli, 1988, Kumbhakar, 1990, Battese and Coelli, 1995 and Filippini and Greene, 2016, among others) assumes that

$$q_i = f(x_i; \beta) \, TE(u_i) \exp(\nu_i), \quad i = 1, 2, \ldots, n, \tag{4.1}$$

where the technical efficiency in production is taken as $TE(u_i) = \exp(-u_i)$, $u_i > 0$, ensuring that this takes values between 0 and 1. Here, q_i is a scalar output, x_i is a $k \times 1$ vector of covariates (a vector of values of known functions of inputs of production and other explanatory variables associated with the i-th firm) and β is a $k \times 1$ vector of parameters to be estimated. As it is usual, the n observations represent a cross-section of firms in a given industry. Furthermore, $\nu_i \in (-\infty, \infty)$ is the noise, and u_i represents technical inefficiency. In the classical literature it is usual to take $\nu_i \sim N(0, \sigma_\nu^2)$, i.e., independently distributed of the u_i and identically distributed as normal distribution with mean zero and variance σ_ν^2. In terms of u_i, various assumptions may be made; for example, Meeusen and Broeck (1977) assigned the exponential distribution to u_i, Battese and Corra (1977) assumed a half normal distribution, while Aigner et al. (1977) considered both distributions. However, since the half normal and exponential distributions are both single-parameter specifications with modes at zero, some scepticism has been expressed regarding their generality. Thus, Stevenson (1980) suggested the truncated normal and gamma distribution for u_i. Greene (1980a,b) proposed the gamma distribution and Lee (1983) proposed a four-parameter Pearson family of distributions. More recently, Greene (1990, 2003) proposed the two-parameter gamma density as a more general alternative in front of the half normal, exponential or truncated normal distributions. Some attempts have been made in order to include some type of dependence between these two random variables. See for instance Smith (2008) and Gómez-Déniz and Pérez-Rodríguez (2015).

We assume here the Cobb-Douglas production function, which is expressed as

$$q(y_1, y_2, \ldots, y_s) = \beta_0 \prod_{i=1}^{s} y_i^{\beta_i},$$

where q is the output or production of the i-th firm, y_i are s inputs quantities of the i-th firm, β_i is a vector of unknown parameters and finally β_0 is a constant which can be interpreted as the part of the output which cannot be explained by the inputs, y_i, usually refereed as the technical change. Then, the model (4.1) can be written as

$$\log q_i = \log \beta_0 + \sum_{i=1}^{s} \beta_i \log y_i + \nu_i - u_i, \quad i = 1, \ldots, n. \tag{4.2}$$

Other functional forms for specifying the production function are possible such as the generalized Leontief, Constant Elasticity of Substitution (CES), etc. (see Coelli et al., 2005, Chap. 8).

Denoting now $\varepsilon_i = \nu_i - u_i$ we have that (4.2) is written as

$$\varepsilon_i = \log q_i - \log \beta_0 - \sum_{i=1}^{s} \beta_i \log y_i, \quad i = 1, 2, \ldots, n. \tag{4.3}$$

In general, the stochastic frontier model (Aigner et al., 1977; Meeusen and Broeck, 1977) in a cross-section framework can be written as $q_i = f(x_i; \beta) + \nu_i \pm u_i$, $i = 1, 2, \ldots, n$, where the sign of the last term depends on whether the frontier describes costs (positive) or production (negative). Thus, the disturbance term, $\varepsilon_i = \nu_i \pm u_i$, which is asymmetric, is assumed to have two components, one with a strictly non-negative distribution, u_i (which is a non-negative component often referred to as the inefficiency term), and another with a symmetric distribution, ν_i (which is termed the idiosyncratic error).

The maximum likelihood method can be used to estimate β, the variances of the errors, and the technical efficiency of each firm. Therefore, distributional assumptions are required for ν_i and u_i.

4.2.1 The general model

We assume the following:

(i) $\nu \sim$ iid $f_1(\nu)$, $-\infty < \nu < \infty$.

(ii) $u \sim$ iid $f_2(u)$, $u > 0$.

(iii) u and ν are distributed independently of each other and of the regressors.

The relevant distributions derived from the assumptions above in the stochastic frontier production (SFP) model, are the marginal distribution of ε obtained from,

$$f(\varepsilon) = \int_0^\infty f_2(u) f_1(u + \varepsilon) \, du, \qquad (4.4)$$

and the conditional distributuion of $u|\varepsilon$ given by

$$f(u|\varepsilon) = \frac{f_2(u) f_1(u + \varepsilon)}{f(\varepsilon)}. \qquad (4.5)$$

Technical efficiency can be computed as

$$\mathrm{TE} = \int_0^\infty \exp(-u) f(u|\varepsilon) \, du. \qquad (4.6)$$

We briefly describe in the next subsections the main models considered in the applied statistical literature in stochastic frontier analysis. They are the normal-half normal, normal-exponential, and normal-truncated normal models.

4.2.2 The normal-exponential model

The classical stochastic frontier model with the normal and exponential assumption (NE) is described by the following stochastic representation:

(i) $\nu_i \sim$ iid $N(0, \sigma_\nu^2)$, $\sigma_\nu > 0$.

(ii) $u_i \sim$ iid exponential with mean $\sigma_u > 0$.

(iii) u_i and ν_i are distributed independently of each other and of the regressors.

The pdf of ν_i is given by

$$f_{\sigma_\nu}(\nu) = \frac{1}{\sigma_\nu \sqrt{2\pi}} \exp\left(-\frac{\nu^2}{2\sigma_\nu^2}\right),$$

$\sigma_u > 0$, while the pdf of u_i is given by,

$$f_{\sigma_u}(u) = \frac{1}{\sigma_u} \exp\left(-\frac{u}{\sigma_u}\right),$$

4.2. STOCHASTIC FRONTIER ANALYSIS

where $-\infty < \nu < \infty$, $\sigma_\nu > 0$, $u > 0$.

In the SPF model, i.e., $\nu = u + \varepsilon$, and by using (4.4) we get,

$$
\begin{aligned}
f_{\sigma_u, \sigma_\nu}(\varepsilon) &= \frac{1}{\sigma_u \sigma_\nu \sqrt{2\pi}} \int_0^\infty \exp\left[-\frac{u}{\sigma_u} - \frac{(u+\varepsilon)^2}{2\sigma_\nu^2}\right] du \\
&= \frac{\exp[-\varepsilon^2/(2\sigma_\nu^2)]}{\sigma_u \sigma_\nu \sqrt{2\pi}} \int_0^\infty \exp\left\{-\frac{1}{2\sigma_\nu^2}\left[u^2 + 2u\left(\varepsilon + \frac{\sigma_\nu^2}{\sigma_u}\right)\right]\right\} du \\
&= \frac{1}{\sigma_u \sigma_\nu \sqrt{2\pi}} \exp\left[\frac{1}{2\sigma_\nu^2}\left(\frac{2\varepsilon\sigma_\nu^2}{\sigma_u} + \frac{\sigma_\nu^4}{\sigma_u^2}\right)\right] \int_{\varepsilon + \frac{\sigma_\nu^2}{\sigma_u}}^\infty \exp\left(-\frac{z^2}{2\sigma_\nu^2}\right) dz \\
&= \frac{1}{\sigma_u} \Phi\left(-\frac{\varepsilon}{\sigma_\nu} - \frac{\sigma_\nu}{\sigma_u}\right) \exp\left(\frac{\varepsilon}{\sigma_u} + \frac{\sigma_\nu^2}{2\sigma_u^2}\right),
\end{aligned} \qquad (4.7)
$$

where $\Phi(\cdot)$ represents the standard normal cdf. The marginal given in (4.7) is asymmetrically distributed with mean $E(\varepsilon) = E(\nu) - E(u) = -\sigma_u$ and variance $var(\varepsilon) = \sigma_u^2 + \sigma_\nu^2$.

Now, after using (4.5) we have that the conditional distribution of u given ε is a truncated normal distribution, $N^+(\widetilde{\mu}, \sigma_\nu^2)$, with pdf given by,

$$
f(u|\varepsilon) = \frac{1}{\sqrt{2\pi}\, \sigma_\nu\, \Phi(\widetilde{\mu}/\sigma_\nu)} \exp\left\{-\frac{1}{2\sigma_\nu^2}(u - \widetilde{\mu})^2\right\}, \qquad (4.8)
$$

where $\widetilde{\mu} = -\varepsilon - \sigma_\nu^2/\sigma_u$.

The mean of the pdf (4.8) is

$$
E(u|\varepsilon) = \widetilde{\mu} + \sigma_\nu \frac{\phi(-\widetilde{\mu}/\sigma_\nu)}{\Phi(\widetilde{\mu}/\sigma_\nu)} = \sigma_\nu\left(\frac{\phi(A)}{\Phi(-A)} - A\right),
$$

being $A = -\widetilde{\mu}/\sigma_\nu$.

Finally, the technical efficiency, computed from (4.8) by using (4.6) is given by

$$
E(e^{-u}|\varepsilon) = \frac{\Phi\left(\widetilde{\mu}/\sigma_\nu - \sigma_\nu\right)}{\Phi\left(\widetilde{\mu}/\sigma_\nu\right)} \exp\left[\varepsilon + \frac{(2 + \sigma_u)\sigma_\nu^2}{2\sigma_u}\right]. \qquad (4.9)
$$

In the stochastic cost frontier (SCF) model, i.e., $v = -u + \varepsilon$, some computations similar to the ones above led us to get the following:

$$
\begin{aligned}
f_{\sigma_u, \sigma_\nu}(\varepsilon) &= \frac{1}{\sigma_u} \Phi\left(\frac{\varepsilon}{\sigma_\nu} - \frac{\sigma_\nu}{\sigma_u}\right) \exp\left\{-\frac{\varepsilon}{\sigma_u} + \frac{\sigma_\nu^2}{2\sigma_u^2}\right\}, \\
f(u|\varepsilon) &= \frac{1}{\sqrt{2\pi}\, \sigma_\nu \Phi(\widetilde{\widetilde{\mu}}/\sigma_\nu)} \exp\left\{-\frac{1}{2\sigma_\nu^2}(u - \widetilde{\widetilde{\mu}})^2\right\},
\end{aligned}
$$

where $\widetilde{\widetilde{\mu}} = \varepsilon - \sigma_\nu^2/\sigma_u$. Again, the marginal $f(\varepsilon)$ is asymmetrically distributed with mean $E(\varepsilon) = \sigma_u$ and variance $var(\varepsilon) = \sigma_u^2 + \sigma_\nu^2$.

4.2.3 The normal-half normal model

We denote this model as NHN. In this case, we assume the following:

(i) $\nu_i \sim$ iid $N(0, \sigma_\nu^2)$.

(ii) $u_i \sim$ iid $N^+(0, \sigma_u^2)$.

(iii) u_i and ν_i are distributed independently of each other and of the regressors.

Here, $N^+(0, \sigma_u^2)$ represents the non-negative half normal, with pdf

$$f_{\sigma_u}(u) = \frac{2}{\sigma_u \sqrt{2\pi}} \exp\left(-\frac{u^2}{2\sigma_u^2}\right), \qquad (4.10)$$

where $u > 0$, $\sigma_\nu > 0$.

Then, in the SPF model, $\nu = u + \varepsilon$, we have that the marginal distribution of ε and the conditional distribution of u given ε are given by

$$f_{\sigma_u, \sigma_\nu}(\varepsilon) = \frac{2}{\sigma} \phi\left(\frac{\varepsilon}{\sigma}\right) \Phi\left(-\frac{\varepsilon \lambda}{\sigma}\right), \qquad (4.11)$$

$$f_{\sigma_u, \sigma_\nu}(u|\varepsilon) = \frac{1}{\sqrt{2\pi}\sigma_*} \left[1 - \Phi\left(-\frac{\mu_*}{\sigma_*}\right)\right]^{-1} \exp\left\{-\frac{(u - \mu_*)^2}{2\sigma_*^2}\right\}, \qquad (4.12)$$

respectively. Here, $\sigma = \sqrt{\sigma_u^2 + \sigma_\nu^2}$, $\lambda = \sigma_u/\sigma_\nu$, $\mu_* = -\varepsilon \sigma_u^2/\sigma^2$, $\sigma_* = \sigma_u \sigma_\nu/\sigma$ and $\Phi(\cdot)$ and $\phi(\cdot)$ are the standard normal cdf and pdf, respectively.

Taking into account the mean and variance of the half-normal with pdf as in (4.10) are given by $E(u) = \sigma_u \sqrt{2/\pi}$ and $var(u) = \sigma_u^2(1 - 2/\pi)$, we get that the mean and variance of the marginal $f(\varepsilon)$ are given by

$$E(\varepsilon) = -\sigma_u \sqrt{\frac{2}{\pi}},$$

$$var(\varepsilon) = \sigma_\nu^2 + \sigma_u^2\left(1 - \frac{2}{\pi}\right),$$

respectively.

Technical efficiency of the ith firm can be calculated from the pdf given in (4.12) by computing $\text{TE}_i = E\left[\exp(-u_i|\varepsilon_i)\right]$, $i = 1, 2, \ldots, n$. and results

$$\text{TE}_i = \frac{1 - \Phi(\sigma_* - \mu_{*i}/\sigma_*)}{1 - \Phi(-\mu_{*i}/\sigma_*)} \exp\left(-\mu_{*i} + \frac{\sigma_*^2}{2}\right), \quad i = 1, 2, \ldots, n, \qquad (4.13)$$

thus, it is calculated using the conditional expectation $E[(\exp(-u_i|\varepsilon_i)]$, conditioned on the composed error $(u_i = \nu_i - \varepsilon_i)$, and evaluated using the computed estimated parameters.

Observe that (4.11) is a reparameterization of the well-known skew-normal distribution described by Azzalini (1985). See also Azzalini (2013, Chap. 3) where the skew normal distribution is connected with stochastic frontier analysis.

4.2.4 The normal-truncated normal model

We assume here a truncated normal (truncated at zero) and normal distributions to describe the inefficiency and the idiosyncratic error, respectively. This model is referred as NTN model and its stochastic representation is given by: (i) $u_i \sim$ iid $TN(\mu, \sigma_u^2)$; (ii) $v_i \sim$ iid normal distribution with parameter $\sigma_v > 0$; and (iii) u_i and v_i are distributed independently of each other and of the regressors. The probability density functions of u_i is as follows (see Stevenson, 1980),

$$f(u) = \frac{1}{\sigma_u \sqrt{2\pi} \bar{\Phi}\left(-\frac{\mu}{\sigma_u}\right)} \exp\left[-\frac{1}{2}\left(\frac{u-\mu}{\sigma_u}\right)^2\right], \quad u > 0,$$

where $\mu \in \mathbb{R}$, $\sigma_u > 0$, $\sigma_v > 0$ and $\bar{\Phi}(\cdot)$ represents the survival function of the standard normal distribution.

Assuming now that the error of the production function is given by $\varepsilon = \nu - u$, the marginal distribution of this error function, assuming independence between u and ν, results

$$f_{\mu,\sigma_u,\sigma_v}(\varepsilon) = \frac{1}{\sigma}\phi\left(\frac{\varepsilon+\mu}{\sigma}\right)\bar{\Phi}\left(\frac{\varepsilon\lambda}{\sigma} - \frac{\mu}{\lambda\sigma}\right)\left[\bar{\Phi}\left(-\frac{\mu}{\sigma_u}\right)\right]^{-1}, \quad -\infty < \varepsilon < \infty, \tag{4.14}$$

where $\lambda = \sigma_u/\sigma_v$ and $\sigma = \sqrt{\sigma_u^2 + \sigma_v^2}$.

The conditional pdf of u given ε is given by

$$f_{\mu,\sigma_u,\sigma_v}(u|\varepsilon) = \frac{1}{\sigma_*}\phi\left(\frac{u}{\sigma_*} - \frac{\mu_*}{\sigma_*}\right)\left[\bar{\Phi}\left(-\frac{\mu_*}{\sigma_*}\right)\right]^{-1},$$

while the technical efficiency obtained from this last pdf results

$$E(e^{-u}|\varepsilon) = \frac{\bar{\Phi}(\sigma_* - \mu_*/\sigma_*)}{\bar{\Phi}(-\mu_*/\sigma_*)} \exp\left(-\mu_* + \frac{1}{2}\sigma_*^2\right), \tag{4.15}$$

where $\mu_* = (\mu\sigma_v^2 - \sigma_u^2\varepsilon)/\sigma^2$ and $\sigma_* = \sigma_u\sigma_v/\sigma$.

Some graphics of the pdf of the marginal distributions (4.11), (4.7) and (4.14) considered are shown in Figure 4.3.

4.2.5 An example

In this section, we estimate the proposed three models studied here by reproducing a study reported by Greene (1980a), who used data originally published in Zellner and Revankar (1969). This data set is reproduced in Table 4.1.

In this study of the transportation-equipment manufacturing industry, observations on added value, capital, and labor are used to estimate a Cobb-Douglas production function.

Table 4.1: Production data (Greene, 1980a).

State	Value add, q	Capital, k	Labor, l
Alabama	126.148	3.804	31.551
California	3201.486	185.446	452.844
Connecticut	690.67	39.712	124.074
Florida	56.296	6.547	19.181
Georgia	304.531	11.53	45.534
Illinois	723.028	58.987	88.391
Indiana	992.169	112.884	148.53
Iowa	35.796	2.698	8.017
Kansas	494.515	10.36	86.189
Kentucky	124.948	5.213	12
Louisiana	73.328	3.763	15.9
Maine	29.467	1.967	6.47
Maryland	415.262	17.546	69.342
Massachusetts	241.53	15.347	39.416
Michigan	4079.554	435.105	490.384
Missouri	652.085	32.84	84.831
New Jersey	667.113	33.292	83.033
New York	940.43	72.974	190.094
Ohio	1611.899	157.978	259.916
Pennsylvania	617.579	34.324	98.152
Texas	527.413	22.736	109.728
Virginia	174.394	7.173	31.301
Washington	636.948	30.807	87.963
West Virginia	22.7	1.543	4.063
Wisconsin	349.711	22.001	52.818

Using (4.7), (4.11) and (4.14) we get the maximum likelihood estimates in the three models considered, which are equal to

$$\ell(\Theta) = n\left(\frac{\sigma_\nu^2}{2\sigma_u^2} - \log\sigma_u\right) + \frac{1}{\sigma_u}\sum_{i=1}^{n}\varepsilon_i + \sum_{i=1}^{n}\log\Phi\left(-\frac{\varepsilon_i}{\sigma_\nu} - \frac{\sigma_\nu}{\sigma_u}\right),$$

$$\ell(\Theta) \propto -n\log\sigma + \sum_{i=1}^{n}\log\Phi\left(-\frac{\lambda\varepsilon_i}{\sigma}\right) - \frac{1}{2\sigma^2}\sum_{i=1}^{n}\varepsilon_i^2,$$

$$\ell(\Theta_1) \propto -n\log\sigma - \frac{1}{2\sigma^2}\sum_{i=1}^{n}(\varepsilon_i+\mu)^2 + \sum_{i=1}^{n}\log\Phi\left(\frac{\varepsilon_i\lambda}{\sigma} - \frac{\mu}{\lambda\sigma}\right)$$
$$-n\log\bar\Phi\left(-\frac{\mu}{\sigma_u}\right),$$

respectively, where $\Theta = (\sigma_u, \sigma_\nu, \beta_0, \beta_1, \beta_2)$ and $\Theta_1 = (\mu, \sigma_u, \sigma_\nu, \beta_0, \beta_1, \beta_2)$ are the vectors of parameters to be estimated and ε_i, from (4.3), is given by

$$\varepsilon_i = \log q_i - \log\beta_0 - \beta_1\log k_i - \beta_2\log l_i, \quad i=1,2,\ldots,n.$$

Table 4.12 shows different estimation methods applied to the data, ordinary least squares (OLS) and maximum likelihood (ML) using both the normal and exponen-

4.2. STOCHASTIC FRONTIER ANALYSIS

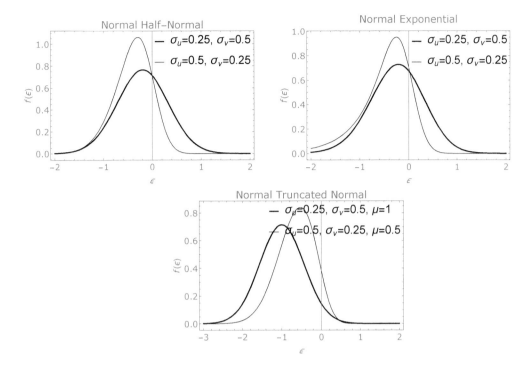

Figure 4.1: Marginal distribution in the NHN, NE and NTN models for different values of parameters.

tial, half normal and truncated normal models. The estimated parameters values and the corresponding maximum value of the log-likelihood function (ℓ_{max}) are shown in this Table.

The OLS estimates of the parameters of the Cobb-Douglas production function can be used as initial values to estimate the maximum likelihood estimates of the parameters. From the analysis, we have observed that the coefficients of capital and labor are statistically significant in the production process. The results indicate that these input variables significantly affect the amount of production in this industry. The estimated parameter is positive, which indicates that the observed output differed from frontier output due to factors within the industries' controls. This implies that the average production function estimated using the OLS was not the proper estimate of the production function in the present case. The intercept value of the ML estimate is greater than the OLS estimate, showing that the estimate of frontier production function lies above the traditional average function.

The results show that the maximum-likelihood estimate of the parameter labor input is about 0.25 for NE and NHN models and about 0.75 for the NTN model. These results also confirm the study of Coelli et al. (2003) where they found that labor has low output elasticity.

All the computations here were performed by using RATS software (see Brooks, 2009).

Table 4.2: Stochastic production frontier estimates.

Variable	Coefficient	t-Stat	Coefficient	t-Stat	Coefficient	t-Stat	Coefficient	t-Stat
	OLS		NE		NHN		NTN	
β_0	1.8444	7.8959	2.0693	7.5993	2.0811	7.7364	2.0785	8.7198
β_1	0.8052	6.3735	0.7704	6.8430	0.7802	6.6608	0.2663	2.8786
β_2	0.2454	2.2968	0.2625	3.3031	0.2585	2.9955	0.7661	6.8657
σ_u			0.1352	2.1663	0.2215	1.8074	0.8907	2.4824
σ_v			0.1714	5.2006	0.1752	4.2840	0.1714	4.3691
μ							-5.4341	3.0520
$\ell_{\max}$	2.2537		2.8605		2.4695		2.8323	

Once the models have been estimated, the technical efficiency (TE) can be obtained by using the maximum likelihood estimates of the parameters and expressions given in (4.13), (4.9) and (4.15). Table 4.13 shows these results, based on the estimation of the parameters shown in Table 4.12 for each model, together with some descriptive statistics for the whole sample.

The mean technical efficiency is found for NE and NTN distributions to be very similar, about 88%. This result shows that the average industry produced only about 88% maximum attainable outputs in this case, whereas it is about 84% maximum output for NHN distribution. In all instances, there is not much variation in the technical efficiencies among the different states.

4.3 Geography: The size distribution of cities

Historically, two approaches have been considered to analyze the empirical distribution of the city size. The Zipf's Law (Zipf, 1949) and the Gibrat's Law of proportionate growth of cities (Gibrat, 1931). The first principle states that the city size follows a power law distribution, i.e., size of larger cities are inversely proportional to their ranks. In this respect, the distribution of the size of urban agglomerations follows a Pareto distribution, or even adheres exactly to the usual Zipf distribution (which is nested in the former model). The Zip's Law in the city distribution system indicates that the second largest city is half the size of the largest, the third largest city a third the size of the largest and the n-th is the n-th size of the largest one. Many papers have argued that the city sizes follow a power law distribution. In this sense, Moura and Ribeiro (2006) studied the Zip's Law for Brazilian cities and Gangopadhyay and Basu (2009)analyzed the size distributions of urban agglomerations for India and China by estimating the scaling exponent for Zip's Law. On the other hand, the Gibrat's Law, perhaps one of the most prominent results in demographics and urban economics, asserts that when the size of the cities grows randomly but proportionately a lognormal distribution is asymptotically generated. The effect of fitting city size data by means of the Pareto distribution vanishes when the whole population range is included in the sample, without excluding medium and small cities. Recently, several papers have shown the superiority of the lognormal distribution over the Pareto, especially when all settlements of a country are examined. In this regard,

4.3. GEOGRAPHY: THE SIZE DISTRIBUTION OF CITIES

Table 4.3: Estimated technical efficiency.

States	Technical efficiency		
	NE	NHH	NTN
Alabama	0.8690	0.8232	0.8662
California	0.9102	0.8693	0.9074
Connecticut	0.8783	0.8318	0.8746
Florida	0.5623	0.6017	0.5630
Georgia	0.9329	0.9041	0.9311
Illinois	0.9226	0.8892	0.9199
Indiana	0.8620	0.8151	0.8568
Iowa	0.8248	0.7854	0.8196
Kansas	0.9354	0.9066	0.9341
Kentucky	0.9600	0.9464	0.9591
Louisiana	0.8651	0.8214	0.8612
Maine	0.8474	0.8061	0.8426
Maryland	0.9143	0.8772	0.9119
Massachusetts	0.8998	0.8596	0.8964
Michigan	0.9013	0.8582	0.8975
Missouri	0.9337	0.9050	0.9318
New Jersey	0.9378	0.9111	0.9360
New York	0.8078	0.7636	0.8028
Ohio	0.8495	0.8010	0.8444
Pennsylvania	0.9052	0.8649	0.9022
Texas	0.8694	0.8217	0.8660
Virginia	0.9106	0.8733	0.9081
Washington	0.9294	0.8984	0.9273
West Virginia	0.8972	0.8603	0.8936
Wisconsin	0.9103	0.8727	0.9074
Mean	0.8815	0.8467	0.8785
Maximum	0.9600	0.9464	0.9591
Minimum	0.5623	0.6016	0.5631
Std. Dev.	0.0765	0.0677	0.0763

Anderson and Ge (2005), also determined that the lognormal model is preferable to the Pareto distribution by using size distribution of Chinese cities. Similarly, Eeckhout (2004) has shown that the lognormal distribution provides a good fit to the size of all cities in the US employing data from 2000 census. However, this argument was objected by Levy (2009) who stated that in the top range of the largest cities, the size distribution diverges dramatically from the lognormal distribution and it is in excellent agreement with a straight line. In this manner, the distribution of the settlement size can be divided into two regions: the bottom and middle ranges where the empirical data are explained by the lognormal distribution, and the top range where the empirical distribution fits a power law distribution. Relying on this idea, Giesen et al. (2010) used the four-parameter Double Pareto lognormal distribution, a distribution that is Pareto in the upper and lower tails and lognormal in between, to explain the distribution of all cities by using untruncated city size data from eight countries. Moreover, other different approaches based on generalization of the

Pareto distribution have also been suggested in the literature. In this regard, Sarabia and Prieto (2009) developed a model to describe Spanish city size data by means of the Pareto Positive Stable distribution. Similarly, Gómez-Déniz and Calderín-Ojeda (2015a) examined the arrangement of urban agglomerations in Australia and New Zealand by using the Pareto ArcTan distribution. In addition, examination of the distribution of city sizes of different countries across a certain period of time has also been considered in the urban economics literature (see Anderson and Ge, 2005). On this subject, Luckstead and Devadoss (2014) studied the city size distribution of China and India for seven decades; in their paper they concluded that the Chinese city distribution is explained by lognormal model between 1950–1990 and by Pareto in 2010. In contrast, the Indian cities fluctuate from lognormal in the earlier periods to Zipf in the most recent periods.

In this section the composite lognormal-Pareto distribution with unrestricted mixing weights is proposed to describe the distribution of the population size of all settlements (*communes*) in France for different years. This issue has been extensively investigated by Calderín-Ojeda (2016). Recent findings have shown that the untruncated settlement size data is in excellent agreement with the lognormal distribution in the lower and central parts of the empirical distribution, but it follows a power law in the upper tail. For that reason, this probabilistic family, that nests in both models, seems appropriate to describe the French commune size data. In this regard, this composite model, that was firstly introduced by Scollnik (2007), uses a lognormal distribution up to an unknown threshold value, estimated from the data, and a two-parameter Pareto density thereafter. Next, continuity and differentiability conditions are imposed at the threshold to yield a smooth density function and to reduce the number of parameters to be estimated. The resulting model is similar in shape to the lognormal distribution but with a thicker tail. Furthermore, numerical results show that for the earlier years, the upper quartile of the commune size distribution is close to Pareto; on the contrary, in the last decade, empirical data are better explained by the lognormal distribution. Composite distributions have been used in actuarial statistics in the last few years to model loss data when the claims faced by insurers consist of a mixture of moderate and large claims. For this kind of data, as the empirical distribution is typically unimodal, highly positively skewed, and includes a thick upper tail, no standard parametric model seems to provide an acceptable fit to both small and large claims, since probabilistic families that provide a good overall fit can perform badly fitting the right tail.

4.3.1 The composite lognormal-Pareto

Scollnik (2007) introduced the three-parameter composite lognormal-Pareto model with unrestricted mixing weights by using adequate truncation of the two-parameter lognormal density up to an unknown threshold value, θ, and the two-parameter Pareto distribution thereafter. This continuous family can be seen then as a convex sum of two density functions in a form of a mixture model. The model that was presented in the Example 1.8 was obtained by imposing continuity and differentiability

4.3. GEOGRAPHY: THE SIZE DISTRIBUTION OF CITIES

conditions at the break point, the scale parameter of the Pareto distribution, not only the resulting density is smooth but also the number of parameters is reduced. Let X be a random variable that follows a composite lognormal-Pareto (CLP) distribution, then its survival function, which will be used later to derive the Zipf plots, is given by

$$S(x) = \begin{cases} 1 - r\Phi\left(\dfrac{\ln x - \mu}{\sigma}\right)\left[\Phi\left(\dfrac{\ln \theta - \mu}{\sigma}\right)\right]^{-1}, & 0 < x \leq \theta, \\ (1-r)\left(\dfrac{\theta}{x}\right)^{\alpha}, & \theta \leq x < \infty. \end{cases}$$

From the latter expression the cdf of the composite lognormal-Pareto model can be easily derived. Besides, as it can be inverted, the quantile function can be simply obtained as follows,

$$Q^{-1}(u) = \begin{cases} \exp\left\{\mu + \sigma\Phi^{-1}\left(\dfrac{u}{r}\Phi\left(\dfrac{\ln \theta - \mu}{\sigma}\right)\right)\right\}, & 0 < u \leq r, \\ \theta\left(\dfrac{1-u}{1-r}\right)^{-1/\alpha}, & r \leq u < 1. \end{cases}$$

Obviously, the inverse transformation method of simulation can be used to generate random variates from the composite lognormal-Pareto distribution.

Let us assume that $\underline{x} = \{x_1, x_2, \ldots, x_k, x_{k+1}, \ldots, x_n\}$ is an ordered random sample selected from the distribution with pdf (1.24). Let us also suppose that the unknown parameter θ satisfies $x_k \leq \theta \leq x_{k+1}$. Then, after writing the unrestricted mixing weight as $r(\mu, \sigma, \theta)$, the log-likelihood function is given by

$$\begin{aligned}\ell(\mu, \sigma, \theta | \underline{x}) &= k\left(\log r(\mu, \sigma, \theta) - \log \Phi\left(\dfrac{\ln \theta - \mu}{\sigma}\right) - \dfrac{1}{2}\log 2\pi - \log \sigma\right) + \sum_{i=1}^{k} \ln x_i \\ &\quad - \dfrac{1}{2}\sum_{i=1}^{k}\left(\dfrac{\ln x_i - \mu}{\sigma}\right)^2 + (n-k)\left(\log(1 - r(\mu, \sigma, \theta)) + \log(\ln \theta - \mu)\right) \\ &\quad - 2\log \sigma + \dfrac{\ln \theta - \mu}{\sigma^2}\log \theta\right) - \left(\dfrac{\ln \theta - \mu}{\sigma^2} + 1\right)\sum_{i=k+1}^{n}\log x_i.\end{aligned}$$

Then, after some tedious algebra, the score equations $\partial \ell/\partial \mu = 0$ and $\partial \ell/\partial \sigma = 0$ are derived. The maximum likelihood estimates of μ and σ are the simultaneous solution of these two equations. Clearly, these estimates cannot be obtained in closed-form and they must be computed numerically. As the log-likelihood function is not continuous with respect to the parameter θ, the maximum likelihood estimate of θ is calculated by segment-wise maximization.

Now, we are investigating the French settlement size data by using the Pareto, lognormal and composite lognormal-Pareto (CLP) distributions.

4.3.2 Data

The *commune* is the fourth-level administrative division of France and it corresponds to the lowest spatial subdivision of the country. Besides, it provides the best coverage since it basically represents the whole French population. Traditionally, French communes are based on pre-existing villages and have been designed with significant power to facilitate local governance. They still largely reflect the division of France into villages at the time of the French Revolution. There is no parallelism between the considerably high number of communes in France than that of any other country. In general, the theories of modelling the city size distribution does not differentiate between an urban agglomeration and a rural one.

The datasets of the commune size distribution have been obtained from the national statistical office (www.insee.fr). In this work communes in the French overseas departments have not been examined. The sets of data considered in this manuscript comprise the estimated population for the years 1962, 1975, 1990, 1999, 2006, and 2012. It is important to point out that only communes with at least one inhabitant have been investigated in this work. Besides, across all the years considered, districts (*arrondissements*) in major cities (Marseille, Lyon, and Paris) have been combined.

Table 4.4: Number of communes and some descriptive statistical measures for the size of the French communes.

Year	Population	Comm.	Mean	S.D.	min	median	max
1962	46425393	36546	1270.3	16531.2	3	356.0	2790091
1975	52591584	36548	1439.0	14852.1	2	333.0	2299830
1990	56615155	36547	1549.1	13956.8	1	365.0	2152423
1999	58518395	36546	1601.2	13950.8	1	380.0	2125246
2006	62817120	36563	1718.1	14587.8	1	420.0	2203817
2012	64844825	36546	1774.3	14943.8	1	443.5	2265886

Some descriptive statistical measures for the empirical distribution of the size of the French communes for the years considered are shown in Table 4.4. As it can be seen the total population of the Metropolitan France has been steadily rising over the last decades. Besides, it can be noticed that that the median is an small number as compared with the mean due to the large proportion of communes with a low population and small proportion of settlements with a large population that makes the empirical distribution of the size of the communes positively skewed. The median shows that the vast majority of the French communes only have a few hundred inhabitants, with a steadily increasing number across the years studied. This fact is confirmed by the relatively large value of the standard deviation. Observe that the latter figure constantly reduces between the years 1962 and 1999 and then, increases again for the four last years under consideration. Note also that the minimum size of the the commune is set equal to one for all the years examined except for 1962,

4.3. GEOGRAPHY: THE SIZE DISTRIBUTION OF CITIES

and 1975. In the following a thorough statistical analysis of these datasets will be carried out by using an untruncated settlement size.

4.3.3 Numerical results

In this subsection, parameter estimation is performed by the method of maximum likelihood (ML), which is implemented using the function "mle"/"mle2" in R, for Pareto, lognormal and composite lognormal-Pareto (CLP) distribution. The ML estimates, across the years considered, for the different models, together with their corresponding standard errors, are reported in Table 4.5, for the distribution of the population size of the French communes after combining districts in major cities.

Table 4.5: Parameter estimates obtained by ML estimation for the models considered for the size of the French communes.

Year	Pareto	lognormal		CLP		
	$\widehat{\alpha}$	$\widehat{\mu}$	$\widehat{\sigma}$	$\widehat{\mu}$	$\widehat{\sigma}$	$\widehat{\theta}$
1962	0.2042	5.9969	1.1506	5.7856	0.8950	782.84
	(0.0011)	(0.0060)	(0.0043)	(0.0070)	(0.0055)	(16.122)
1975	0.1680	5.9529	1.2708	5.6972	0.9738	746.21
	(0.0009)	(0.0066)	(0.0047)	(0.0061)	(0.0048)	(11.115)
1990	0.1655	6.0415	1.3241	5.8099	1.0601	969.10
	(0.0009)	(0.0069)	(0.0049)	(0.0063)	(0.0049)	(15.185)
1999	0.1646	6.0770	1.3345	5.8357	1.0655	988.04
	(0.0009)	(0.0070)	(0.0049)	(0.0061)	(0.0047)	(14.096)
2006	0.1621	6.1688	1.3348	5.9598	1.0965	1232.82
	(0.0008)	(0.0070)	(0.0049)	(0.0105)	(0.0084)	(46.188)
2012	0.1609	6.2161	1.3375	6.0222	1.1161	1393.32
	(0.0008)	(0.0070)	(0.0049)	(0.0067)	(0.0052)	(26.248)

As it can be observed in this Table, the estimate of the Pareto exponent is well below one, likewise departing from the Zipf's Law; besides its value is steadily decreasing ranging from 0.2042 in the year 1962 to 0.1609 in the year 2012. Moreover, the location parameter of the lognormal distribution μ, firstly decreases in the earlier years, and then it increases over time, and the scale parameter σ steadily grows in the period of investigation. Therefore, both the mean and variance of the lognormal distribution increase over time showing not only the population growth in the communes but also that differences among the size of the settlements widen. Finally the parameters of the CLP model consistently rise across the years considered, indicating the prevalence of the lognormal distribution over the Pareto distribution in the composite model with time. This is confirmed by the values of the unrestricted mixing weight r as it is displayed in Table 4.6. In this sense, for example, the distribution of the French communes for 1975 follows a power law distribution from percentile 75 onwards, similarly, for the year 2012, this occurs beyond percentile 81. For the former case, settlements with a population between 1 and 749 inhabitants grows randomly and proportionately, providing a consistent picture of the commune size

Table 4.6: Values of tail index α and unrestricted mixing weight r for the size of the French communes.

	1962	1975	1990	1999	2006	2012
$\hat{\alpha}$	1.0953	0.9678	0.9495	0.9338	0.9626	0.9772
r	0.7687	0.7528	0.7796	0.7746	0.7978	0.8104

dynamics; then cities of medium-to moderate size are better represented by a power law. On the other hand, for the latter year considered, the distribution of cities between 1 and 1475 inhabitants adopt the pure structure of the Gibrat's Law, again the growth of cities of larger size are better described by the Pareto distribution. These results are supported by the Zip's plots (see Figure 4.2). In a similar fashion, the tail index α is very close to unity for all the years considered, satisfying therefore the Zipf's Law. However, one need to be cautious by comparing these values with the existing coefficients of the Zip's Law s in the literature since they are highly sensitive to the chosen threshold (e.g., minimum value) city (Giesen et al., 2010).

Model assessment is presented from a theoretical plausibility justified by means of Kullback-Leibler divergence, suggesting using an information-criterion based approach. In application, NLL and HQIC have been chosen as measures of model validation. Note that for these two information criterion described above, smaller values indicate a better fit of the model to the data. As it can be seen in the results shown in Table 4.7, the lognormal-Pareto family and outperforms the Pareto and lognormal distributions consistently across the different years analyzed in this work for French population data.

Table 4.7: NLL (above) and HQIC (below) values evaluated at ML estimates of the models considered for the size of the French communes.

Year	Pareto	lognormal	CLP
1962	313775	276146	274284
	627560	552307	548587
1975	319311	278184	276410
	638630	556382	552840
1990	323081	282917	281695
	646171	565848	563408
1999	324584	284492	283363
	649177	568999	566744
2006	328638	287987	286994
	657285	575987	574006
2012	330492	289657	288776
	660994	579328	577570

In the following, model validation is also presented by using a measure of goodness-of-fit based on the empirical distribution function to quantify this distance between the empirical distribution function constructed from the data and the cumulative distribution function of the fitted models, the KS test statistic. For this model as-

4.3. GEOGRAPHY: THE SIZE DISTRIBUTION OF CITIES

Table 4.8: Kolmogorov-Smirnov test statistic (KS) and its corresponding p-values (in brackets) for Pareto, lognormal and CLP distributions for the size of the French communes.

Year	Pareto	lognormal	CLP
1962	0.4387 (0.000)	0.0547 (0.000)	0.0073 (0.038)
1975	0.3932 (0.000)	0.0576 (0.000)	0.0084 (0.012)
1990	0.4477 (0.000)	0.0478 (0.000)	0.0112 (0.000)
1999	0.4473 (0.000)	0.0452 (0.000)	0.0114 (0.000)
2006	0.4493 (0.000)	0.0424 (0.000)	0.0129 (0.000)
2012	0.4482 (0.000)	0.0404 (0.000)	0.0132 (0.000)

sessment measure, a smaller value indicates a better fit of the distribution to the empirical data. Results are summarized in Table 4.8.

The KS test statistic is clearly much larger for the Pareto distribution than for the other two models. By comparing the lognormal and CLP distributions, the values of the test statistic are greater for the former model than for the latter one consistently across all the years considered. However, the numerical value of the test statistic declines with time for the lognormal distribution whereas for the CLP model it increases. The test statistic also allows us to perform hypothesis testing for model validation purposes if it is assumed that all parameters are specified completely. An extremely small p-value may lead to a confident rejection of the null hypothesis that the data come from the proposed model. Observe that the p-value is lower than 0.001 in all cases except for the years 1962, and 1975 in the CLP model. However, it is relevant to mention that the KS-test would reject the Pareto and lognormal distribution earlier than the CLP model for all the years examined. Although, it is not relevant for this book, the p-values were computed via Monte Carlo methods using a simulation size of 10000 repetitions.

The linear relationship between the population of the settlements and their corresponding ranks on a log-log plot (Zipf plots) is found to be a power law, where the absolute value of this linear expression is the exponent of the power law. Figure 4.2 illustrates Zipf plots of actual and predicted values of the three distributions for the sample cities of the French communes for the years 1962, 1975, 1990, 1999, 2006, and 2012 respectively. These two sets of graphical representation correspond to the plots, in log-log scale, of the complementary of the cdf against the observed ordered data for Pareto, lognormal and composite lognormal-Pareto (CLP) distributions together with the empirical distribution. Logarithmic of empirical quantiles appears as scatter points on the chart, the logarithm of the theoretical quantiles are given by lines: Pareto (dotted), lognormal (dashed), and CLP (solid). Note that for the Pareto case a straight line with slope $1/\alpha$ is obtained. Observe that for all the years considered the Pareto distribution does not perform well. Although the lognormal distribution does not explain the empirical data in the years considered, it can be observed that the fit to data improve slightly in the second set of years (see bottom row of Figure 4.2), likewise corroborating the results given in Table 4.6. As it can be seen, the CLP distribution stays closer to the empirical quantiles across all the years considered. This effect is even more evident in the years 1962, and 1975. Thus, it can be concluded that, for these three years, the upper quartile of the settlement size distribution adheres closely to a power law distribution. Then, as time goes by, this effect vanishes. In particular for the last decade, although the French population has been steadily increasing, only the range covering communes from medium-to-large size are described by the Zip's Law. For these years (see middle and bottom row of Figure 4.2), the Pareto distribution tends to overestimate the empirical distribution for the communes of large-to-extreme size. These results are also supported by the value of the mixing weight r. Note that for the lasts years, the lognormal distribution explains more than 80% of the composite model.

4.4 ACD models

Autoregressive conditional duration (ACD) models deal with statistical tools for the analysis of data that arrive at irregular intervals. These models treat the time between events as a stochastic process with dependent arrival rates focusing on the expected duration between events. As Bhatti (2009) points out, *every trading day is different, differing in its number of aggregate trades, the total volume traded, and daily price movement.*

The duration between events in economic agents has been widely considered in theoretical and empirical studies since 1970 (see for instance Pyrlik, 2013). However, the seminal paper in modern form is due to Engle and Russell (1998). The survival analysis generally assumes independence in the events studied, and consequently, the durations between events are independent.[1] The classical model is defined as follows. Let a stochastic process that is simply a sequence of times $\{T_0, T_1, \ldots, T_n, \ldots\}$ with $T_0 < T_1 < \cdots < T_n < \ldots$ denote a sequence of random transaction times and define the ith trade duration as $X_i = T_i - T_{i-1}$, which represents the duration between consecutive points. This interval between two arrival times will be called the duration. The main idea of the ACD model is that the temporal dynamics of the trade durations can be summarized by their mean behavior. Let $\psi_i = E(X_i|\mathcal{F}_{i-1})$, where $\mathcal{F}_{i-1} = \{x_1, \ldots, x_{i-1}\}$, denotes the conditional mean of the ith trade duration based on the past behavior of the point process up to time T_{i-1} denoted by $\mathcal{F}_{i-1}$.

This represents the information set available at time T_{i-1}. The ACD model introduced by Engle and Russell (2006) assumes that

$$X_i = \psi_i \varepsilon_i,$$

where ε_i is a unit mean positive random variable, i.e., $E(\varepsilon_i) = 1$, and ψ_i and ε_i are stochastically independent. That is, $E(\psi_i \varepsilon_i) = E(\psi_i) E(\varepsilon_i)$. Since the conditional expectation of the duration depends upon past durations the model is named as autoregressive conditional duration (ACD). The assumption of that $E(\varepsilon_i) = 1$ can be relaxed by defining $\varepsilon_i^* = \varepsilon_i / E(\varepsilon_i)$ and $\psi_i^* = \psi_i E(\varepsilon_i)$. Thus, in this case we get that $E(\varepsilon_i^*) = 1$ and $X_i = \psi_i^* \varepsilon_i^*$.

---

[1] However, Engle and Russell (2006) introduced a model in which both the events studied, such as the duration between them, are interdependent temporarily.

4.4. ACD MODELS

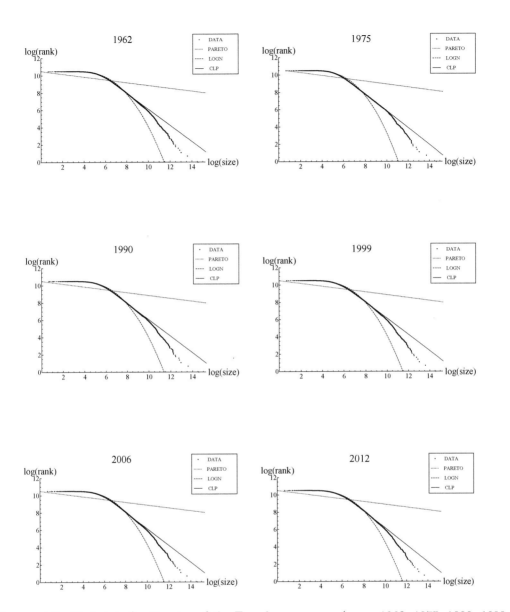

Figure 4.2: Zipf plots for the size of the French communes (years 1962, 1975, 1990, 1999, 2006 and 2012).

The general formulation can be specified formally as follows (see Luca and Gallo, 2009):

$$\begin{aligned}
X_i &= \phi(t_i)\psi_i\varepsilon_i, \\
\psi_i &= f(x_{i-1},\ldots,x_{i-q},\psi_{i-1},\ldots,\psi_{i-p}) \\
&= \omega + \sum_{j=1}^{q}\alpha_j x_{i-j} + \sum_{j=1}^{p}\beta_j\psi_{i-j}, \quad \omega > 0,\ \alpha_j \geq 0,\ \beta_j \geq 0,
\end{aligned} \quad (4.16)$$

where ε_i are iid with positive support, $E(\varepsilon_i) = 1$ and $f : \mathbb{R}_+ \longrightarrow \mathbb{R}_+$. This is a special case of the more general class of Multiplicative Error Model (MEM) contemplated by Engle and Gallo (2006). The subscript i refers to the ith market event recorded at time t_i, $X_i = T_i - T_{i-1}$ and $\phi(t_i)$ is a daily seasonal component. By taking now $x_i = X_i/\phi(t_i)$ such that x_i is the seasonally adjusted duration and denoting the information set up to time t_{i-1} as $\mathcal{F}_{i-1}$, in the ACD model results $E(x_i|\mathcal{F}_{i-1}) = \psi_i$. Thus, ψ_i is the expected duration conditionally on the information up to the time t_{i-1} and is conditionally deterministic.

We have that under the ACD(1,1) model, the most simple, the mean and variance are given by

$$E(X_i) = \eta = \frac{\omega}{1 - \alpha - \beta},$$
$$var(X_i) = \eta^2 \left(\frac{1 - \beta^2 - \alpha\beta}{1 - \beta^2 - 2\alpha\beta - 2\alpha^2} \right).$$

The simplest distributional assumptions for conditional excess durations is the exponential. But, this pdf is far from capturing some features of the errors, such as their variability. Thus, alternative pdfs have been proposed in the vast statistical literature. Some of them are the gamma, Weibull, the generalized gamma proposed by Stacy (1962), the Burr, lognormal, Pareto and Birnbaum-Saunders distributions. See for details Lunde (1999); Engle and Russell (2006); Luca and Gallo (2009) and Bhatti, 2010; among others.

4.4.1 The general model

Assuming that $X \sim f(x)$ with finite expectation $E(X) = \mu > 0$ and writing $\varepsilon_i = X/\mu$ we have that $E(\varepsilon_i) = 1$. Now, if $X = \mu\varepsilon_i$ we get that $f_\varepsilon(\varepsilon_i) = \mu f(\mu\varepsilon_i)$. Therefore we finally obtain

$$f_{X/\psi}(x_i|\psi_i) = \frac{\mu}{\psi_i} f(\mu x_i/\psi_i), \quad i = 2, \ldots, t. \tag{4.17}$$

The parameters of the model are estimated by maximizing the recursively defined log-likelihood function with increments given by

$$\ell(\tilde{x}; \Theta) = \log \mu - \log \psi_i + \log f(\mu x_i/\psi_i), \tag{4.18}$$

where ψ_i should be replaced by (4.16), $\tilde{x} = \{x_1, \ldots, x_t\}$ is the sample mean and Θ is the set of parameters which has to be estimated, α, β, ω and the remaining parameters of the pdf $f(\cdot)$.

4.4.2 Specific models

Some particular choices of the pdf $f(x)$ are considered now in the following sections.

The exponential ACD model

In this case, it is assumed that X follows an exponential distribution with mean $\theta > 0$. Thus, by using (4.17) we get

$$f_{X/\psi}(x_i|\psi_i) = \frac{1}{\psi_i} \exp\left\{-\frac{x_i}{\psi_i}\right\},$$

and assuming an ACD(1,1), i.e.,

$$\psi_i = \omega + \alpha\, x_{i-1} + \beta\, \psi_{i-1}$$

we have that by using (4.18), the log-likelihood function results,

$$\ell(\tilde{x}; \Theta) = \log f(x_i|\psi_i) = -\sum_{i=1}^{t} \left[\log \psi_i + \frac{x_i}{\psi_i}\right], \quad \Theta = (\alpha, \beta, \omega),$$

with $\psi_1 = \omega/(1-\beta)$ (see Engle and Russell, 2006).

Following code, developed in Mathematica v.12, provides the estimation, standard errors, t-Statistics, p-values of the exponencial ACD(1,1) model and diagnostic of the same by giving a table with the maximum value of the log-likelihood function, the AIC, BIC, and CAIC(Bozdogan , 1987). The data can be read in any form as a list.

```
Clear[w,alfa,beta];
k =Length[data];Array[X,k];Array[m,k];
m[1]=w/(1-beta);Do[X[s]=data[[s]],{s,1,k}];
Do[m[l]=w+alpha*X[l-1]+beta*m[l-1],{l,2,k}];
logl=Table[-Log[m[p]]-X[p]/m[p],{p,1,k}];
loglike=Sum[logl[[j]],{j,1,k}];
maxlogl=FindMaximum[loglike,{w,0.18,.2},{alpha,0.051,0.7},
    {beta,0.8,0.9}]
mle=maxlogl[[2]];parameters={w,alpha,beta};
f=Experimental`CreateNumericalFunction[parameters,
    loglike,{1},Hessian->FiniteDifference];
MH=Flatten[-f["Hessian"[parameters/.mle]],1];
B=Inverse[MH];param=Length[parameters];
erroreparam=Table[N[Sqrt[B[[j,j]]]],{j,1,param}];
t=Table[N[parameters[[j]]/erroreparam[[j]]]/.mle,{j,1,param}];
pvalues=Table[2*(1-CDF[StudentTDistribution[k-param],
    Abs[t[[j]]]])/.mle,{j,1,param}];
AIC=N[2*param-2*loglike/.mle];
BIC=N[-2*(loglike/.mle)+param*Log[k]];
CAIC=N[(1+Log[k])*param-2*loglike/.mle];
listaestimate=parameters/.mle;
listase=erroreparam;listat=t;
```

```
listapvalue=pvalues;listparameters=parameters;
MatrixForm[Transpose[{listparameters,listaestimate,
   listase,Abs[listat],listapvalue}]];
TableForm[Table[{listparameters,listaestimate,listase,
   Abs[listat],listapvalue},{i,1,1}],TableHeadings->{None,
   {"Parameter","Estimate","S.E.","t-Wald","P>|t|"}},
   TableSpacing->{1,3}]
TableForm[Table[{loglike/.mle,AIC,BIC,CAIC},{i,1,1}],
   TableHeadings->{None,{"Log-likelihood","AIC","BIC","CAIC"}},
   TableSpacing->{1,3}]
```

The Weibull ACD model

A continuous and positive random variable X follows a Weibull distribution if its pdf is given by

$$f(x) = \theta x^{\theta-1} \exp\left(-x^\theta\right), \quad x > 0, \, \theta > 0.$$

The mean of this distribution is $E(X) = \Gamma(1/\theta)/\theta$. Now, by writing $\phi(\theta) = \theta/\Gamma(1/\theta)$ and using (4.18) we get the log-likelihood of the Weibull ACD model given by,

$$\begin{aligned}\ell(\tilde{x};\Theta) &= t\log\theta + (\theta-1)\sum_{i=1}^{t}\log x_i - \theta\left(\log\phi(\theta) + \sum_{i=1}^{t}\log\psi_i\right)\\ &\quad - \sum_{i=1}^{t}\left(\frac{x_i}{\psi_i\,\phi(\theta)}\right)^\theta, \quad \Theta = (\theta,\omega,\alpha,\beta).\end{aligned}$$

The generalized exponential ACD model

A continuous random variable X follows the generalized exponential distribution proposed by Marshall and Olkin (1997) if its pdf is given by

$$f(x) = \frac{\gamma e^{-x/\theta}}{\theta(1-\bar{\gamma}e^{-x/\theta})^2}, \qquad (4.19)$$

for $x > 0$, $\theta > 0$, $\gamma > 0$ and $\bar{\gamma} = 1 - \gamma$. The particular case $\gamma = 1$ reduces to the exponential distribution with mean $\theta > 0$, studied in Section 1.3.4. The mean of this distribution is given by $\mu = -\theta(\gamma/\bar{\gamma})\log\gamma$. Again, by using (4.18) together with (4.19) we get the log-likelihood, given by

$$\begin{aligned}\ell(\tilde{x};\Theta) &= t\left[\log(\phi(\gamma)) + \log\gamma\right] - \sum_{i=1}^{t}\log\psi_i - \phi(\gamma)\sum_{i=1}^{t}\frac{x_i}{\psi_i}\\ &\quad -2\sum_{i=1}^{t}\log\left[1-\bar{\gamma}\exp\left(-\frac{\phi(\gamma)x_i}{\psi_i}\right)\right], \quad \Theta = (\omega,\alpha,\beta,\gamma).\end{aligned}$$

where $\phi(\gamma) = -(\gamma/\bar{\gamma})\log\gamma$.

4.4. ACD MODELS

The Burr ACD model

A continuous random variable X follows the Burr distribution if its pdf is given by

$$f(x) = \frac{\gamma \theta x^{\gamma-1}}{(1+x^\gamma)^{\theta+1}} \qquad (4.20)$$

for $x > 0$, $\theta > 0$, $\gamma > 0$. The mean of this distribution is given by $\mu = \phi(\theta, \gamma) = \theta B(\theta - 1/\gamma, 1 + 1/\gamma)$, where $B(x,y) = \int_0^1 z^{x-1}(1-z)^{y-1} dz$ represents the beta function. Now, by using (4.18) and (4.20) we get the log-likelihood for the Burr ACD model given by,

$$\begin{aligned}\ell(\tilde{x}; \Theta) &= t[\log\theta + \log\gamma] - \sum_{i=1}^{t}\log x_i + \gamma\sum_{i=1}^{t}\log\left(\frac{\phi(\theta,\gamma)x_i}{\psi_i}\right) \\ &\quad -(\theta+1)\sum_{i=1}^{t}\log\left[1+\left(\frac{\phi(\theta,\gamma)x_i}{\psi_i}\right)^\gamma\right], \quad \Theta=(\omega,\alpha,\beta,\gamma,\theta).\end{aligned}$$

The inverse Gaussian ACD model

The inverse Gaussian distribution with pdf given in (1.14) is considered now. Recall that a continuous random variable X follows the inverse Gaussian distribution if its pdf is given by

$$f(x) = \sqrt{\frac{\theta}{2\pi}} x^{-3/2} \exp\left[\frac{\theta}{\gamma} - \frac{\theta}{2}\left(\frac{x}{\gamma^2} + \frac{1}{x}\right)\right], \qquad (4.21)$$

for $x > 0$, $\theta > 0$, $\gamma > 0$. The mean of this distribution is given by $\mu = \gamma$. In this case, after using (4.18) together with (4.21) we have that the log-likelihood is given by

$$\begin{aligned}\ell(\tilde{x}; \Theta) &= \frac{t}{2}[\log\theta - \log(2\pi) - \log\gamma] + \frac{1}{2}\sum_{i=1}^{t}\log\psi_i \\ &\quad + \frac{t\theta}{\gamma} - \frac{3}{2}\sum_{i=1}^{t}\log x_i - \frac{\theta}{2\gamma}\sum_{i=1}^{t}\left(\frac{x_i}{\psi_i}+\frac{\psi_i}{x_i}\right),\end{aligned} \qquad (4.22)$$

where $\Theta = (\omega, \alpha, \beta, \theta, \gamma)$.

A more simple model can be computed by assuming $\gamma^2 = \theta$, and in this case, the log-likelihood in (4.22) depends on one parameter.

4.4.3 Extensions

The most recent focus of attention concerning the financial durations of transactions has been the unobserved heterogeneity that can be caused, for example, by differences in trading conditions, and which are not readily captured by covariates (observed heterogeneity). It is well known that duration analysis produces incorrect results if unobserved heterogeneity is ignored because this can have serious consequences on the estimation of parameters, which can be sensitive to the presence of unobserved heterogeneity. The link between statistical and financial aspects within a set of distributional assumptions is based on financial market microstructure theories. These theories divide traders into informed and non-informed, and the distribution of duration is assumed to be derived from a mixture of distributions. This basic idea is not new in micro econometric studies (see for instance, Lancaster, 1990). The heterogeneity affects the hazard model in a sense that for traders who belong to distinct categories, durations might obey different probability laws. In fact, the assumption of interaction among agents, informed traders who possess private information and liquidity traders whose the information set is publicly available, O'Hara (1995) suggests that financial durations may obey different probability laws. On the other hand, there are many reasons to believe that arrival rates for informed and uninformed agents exhibit temporal dependence, each with its distinct pattern. See also Honoré (1990), Abbring and van den Berg (2007), Cho and White (2010), Gómez-Déniz and Pérez-Rodríguez (2016) and Gómez-Déniz and Pérez-Rodríguez (2017).

Authors such as Luca and Zuccolotto (2003), Luca and Gallo (2004) and Luca and Gallo (2009) have proposed various mixtures of distributions to specify and estimate unobserved heterogeneity in the context of ACD models. The simplest formulation is the mixture of exponential distributions, which can be a finite mixture, if the traders are assumed to be divided into a finite number of groups, or an infinite mixture when every trader is considered to have their behavior. Therefore, the infinite mixture summarises a wide variety of agents or trading conditions. Thus, different degrees of information and different attitudes toward risk, budget constraints, and so on can be taken into account, allowing for a complex unobserved heterogeneity. This context translates into a mixture provided the assumption holds of no interaction among agents: relaxing this hypothesis complicates matters and could be the object of further study. If test results indicate unobserved heterogeneity, then the crucial issue is to incorporate the mixing distribution of the heterogeneity term.

Because the parameter estimates are very sensitive to the choice of the mixing distribution (Heckman and Singer, 1984), other mixing distributions have sometimes been considered. Although there is no an argument in favor of one choice over the other, we consider a distribution which can accommodate two facts: that the intensity function conditional on past durations (hazard function) is non-monotonic, and that the unobserved heterogeneity of traders can be modeled using finite mixtures of non-exponentials.

Our attention here is paid to the mixture of the exponential distribution with the inverse Gaussian one.

4.4. ACD MODELS

The exponential inverse Gaussian distribution (Bhattacharya and Kumar, 1986 and Frangos and Karlis, 2004) studied in Chapter 1 is obtained by mixing an exponential distribution with mean $\lambda > 0$ with the inverse Gaussian distribution (4.21). The pdf of the exponential-inverse Gaussian distribution is given by:

$$f(x) = \frac{1}{\gamma} \left[\frac{\theta}{\gamma \delta(x;\theta)} \right]^2 \left[1 + \frac{1}{\delta(x;\theta)} \right] \exp\left[\theta/\gamma - \delta(x;\theta)\right], \qquad (4.23)$$

where

$$\delta(x;\theta) = \frac{1}{\gamma}\sqrt{\theta(\theta + 2x)},$$

for $x > 0$, $\theta > 0$ and $\gamma > 0$.

The mean of this distribution is $\mu = \gamma$, and the log-likelihood function for the exponential-inverse Gaussian autoregressive conditional duration model is obtained from (4.18) and (4.23), giving,

$$\begin{aligned}\ell(\tilde{x};\Theta) &= 2t(\log\theta - \log\gamma) - \sum_{i=1}^{t}\log\psi_i - 2\sum_{i=1}^{t}\log\delta(\gamma x_i/\psi_i) \\ &\quad - \sum_{i=1}^{t}\delta(\gamma x_i/\psi_i) + \sum_{i=1}^{t}\log\left[1 + \frac{1}{\delta(\gamma x_i/\psi_i)}\right], \quad \Theta = (\omega,\alpha,\beta,\theta,\gamma).\end{aligned}$$

4.4.4 An empirical example

As an illustration of the application of our specification to financial duration data, we estimated a simple ACD(1,1) model using data obtained from the transaction durations of IBM stock on five consecutive trading days from November 1 to November 7, 1990, adjusted by removing the deterministic component (Tsay, 2002). The number of observations employed was restricted to 2000 positive adjusted durations. A sample of the 100 first observations together with some descriptive statistics are displayed in Table 4.9. Table 4.10 summarises the results of different distributions used in this paper to estimate the ACD(1,1) model. We consider the exponential, Weibull, generalized gamma, Burr, inverse Gaussian with two and one-parameter distributions. Let Θ be a vector of unknown parameters to be estimated. Then, the density functions and the logarithm of the different duration models used in this paper are as follows:

Because the tests for ACD models involve basic residual examinations, testing the functional form of the conditional mean duration or testing the distribution of the error term, we use some statistics based on standardized durations: mean, standard

Table 4.9: 100 first observations for ACD model.

2.5868	1.5258	9.6265	1.7398
0.3233	3.4237	0.5921	0.8024
1.6151	1.8992	3.9817	3.3331
1.2913	3.0313	12.8221	5.8353
19.8153	7.1582	29.3389	0.6627
0.6390	3.0512	0.7082	3.8304
3.8266	1.1229	14.5327	43.422
6.9907	2.8797	0.6982	3.6701
1.2702	1.5975	5.2808	6.6681
4.7519	2.2318	3.1869	0.3773
16.9604	11.1996	0.4155	0.3772
3.7618	0.1577	1.6595	0.5027
3.7547	2.3605	0.2765	4.0075
6.8597	4.0760	0.8289	6.7232
3.1131	2.8144	1.2420	3.9704
1.2445	5.2897	3.7132	0.7440
5.8936	2.4835	0.4124	9.2257
4.1240	4.4824	3.5626	1.4742
4.6955	4.0037	3.2787	1.2272
9.8991	2.1515	0.2732	1.7156
6.1806	3.9807	8.5373	5.1243
0.7721	1.2234	0.6771	2.9209
1.9273	4.5683	7.3983	3.7608
9.3729	17.6714	2.1480	3.2665
2.1006	2.2417	1.7425	0.1210

Mean	3.29178
Maximum	43.422
Minimum	0.079
Std. Dev.	4.07558

deviation, excess-dispersion,

$$E(x_i|\psi_i) = \frac{1}{t}\sum_{i=1}^{t}\frac{x_i}{\psi_i},$$

$$\sigma_{x_i|\psi_i} = \sqrt{\frac{1}{t}\sum_{i=1}^{t}\left(\frac{x_i}{\psi_i} - E(x_i|\psi_i)\right)^2},$$

$$\text{MSE}_{x_i|\psi_i} = \frac{1}{t}\sum_{i=1}^{t}(x_i - \psi_i)^2,$$

$$\text{MAE}_{x_i|\psi_i} = \frac{1}{t}\sum_{i=1}^{t}|x_i - \psi_i|,$$

$$\text{Excess-Dispersion} = \sqrt{t}\frac{\sigma_{x_i|\psi_i}^2 - 1}{\sigma_{(x_i|\psi_i-1)^2}}.$$

Akaike and Schwarz Bayesian information criteria (AIC and BIC, respectively), durations ($\text{MSE}_{x_i|\psi_i}$, $\text{MAE}_{x_i|\psi_i}$), and also the Ljung-Box statistic ($Q(k)$) for $k = 2, 5, 10$ and 20 lags for autocorrelation in the standardized residuals are shown in Table 4.10.

Table 4.10 summarises the results of different distributions used in this book to estimate the ACD$(1,1)$ model. We consider the exponential, Weibull, generalised gamma, Burr, inverse Gaussian and the mixture exponential-inverse Gaussian distribution.

In general, these results indicate that all parameters in each estimated model are statistically significant at any conventional level of significance. Nevertheless, the standardized residual does not present serial dependence. Henceforth, misspecification tests based on the autocorrelation of standardized residuals indicate the non-presence of serial dependence. All the fitted models present good behavior in their standardized residuals, so they cannot be compared using this portmanteau test. On the other hand, in terms of the log-likelihood measures and AIC and BIC information criteria, inverse Gaussian fits better than other distributions. Moreover, concerning statistical error measures, the inverse Gaussian has a lower MSE than do other models. Finally, as can be seen, the durations are overdispersed in all models because the mean value is lower than the variance.

4.5 Income

The analysis of income distribution has a venerable history in economics and statistics. Over the last decades, there has been a growing interest in its evolution in the economic literature and the international policy fora. The Lorenz curve (henceforth, LC) is a central instrument for the study of the distribution of income, which informs about the cumulative proportion of income held by the bottom p percent of the population. Its graphical representation offers a complete picture of the concentration in the distribution and its popularity has spread to fields beyond economics, such as bibliometrics and physics (Burrell, 1991, Burrell, 2006, Egghe, 2005 and Rousseau, 2007).

The fundamental role of these curves for the analysis of inequality makes the development of new functional forms particularly relevant. Gastwirth (1971) introduced a formal definition of the LC that admitted a parametric representation. This approach has motivated a substantial body of research in this area. After the proposal of Kakwani and Podder (1976), numerous alternatives have been developed, among which we highlight Rasche et al. (1980), Aggarwal (1984), Aggarwal and Singh (1984), Gupta (1984), Chotikapanich (1993), Ryu and Slottje (1996).[2] More recent proposals are Ryu and Slottje (1996), Sarabia et al. (1999), Sarabia et al. (2010a), Sarabia et al. (2019) and Gómez-Déniz (2016).

However, estimating a parametric LC instead of a parametric income distribution might present potential limitations. First, the underlying income distribution derived

---

[2] For a detailed reading about the LC concept, we refer the reader to Chotikapanich (2008).

Table 4.10: Maximum likelihood estimates, statistics and misspecification tests of the different ACD(1,1) models.

	Exponential	Weibull	Generalised Gamma	Burr	Inverse Gaussian	Generalized exponential	Exponential Inverse Gaussian
γ		0.8367	4.6405	0.9809	1.7309	0.4435	0.0327
		(0.0205)	(1.2551)	(0.0213)	(0.0417)	(0.0429)	(0.0028)
θ			0.3742	4.9422	0.4593		0.0910
			(0.0537)	(1.0121)	(0.0093)		(0.0053)
ω	0.1121	0.1121	0.0894	0.0913	0.1027	0.0994	0.0914
	(0.059)	(0.0799)	(0.0411)	(0.0414)	(0.0444)	(0.0429)	(0.0373)
α	0.0688	0.0688	0.0676	0.0681	0.0491	0.0689	0.0684
	(0.015)	(0.0161)	(0.0123)	(0.0127)	(0.0106)	(0.0199)	(0.0102)
β	0.8979	0.8979	0.9077	0.9059	0.9195	0.9019	0.9055
	(0.0282)	(0.0343)	(0.0189)	(0.0199)	(0.0194)	(0.0259)	(0.0166)
NLL	4310.737	4288.5227	4242.4724	4262.5508	4366.9023	4257.1636	4259.566
AIC	8627.470	8585.050	8494.940	8535.100	8743.800	8522.330	8529.130
BIC	8644.280	8607.450	8522.950	8563.100	8771.810	8544.730	8557.130
CAIC	8647.28	8611.450	8527.950	8568.100	8776.810	8548.730	8562.130
$E(x_i|\psi_i)$	1.0059	1.0052	0.9905	0.9986	1.008	1.0059	0.9994
$\sigma_{x_i|\psi_i}$	1.2631	1.2631	1.2482	1.2577	1.2696	1.2654	1.2586
$\mathrm{MSE}_{x_i|\psi_i}$	16.747	16.748	16.7745	16.7677	16.6922	16.7581	16.7684
$\mathrm{MAE}_{x_i|\psi_i}$	2.7331	2.7331	2.7539	2.7448	2.7211	2.7353	2.7442
Excess-Dispersion	3.4389	3.4391	3.2043	3.3092	3.4415	3.4181	3.3195
$Q(1)$	0.266	0.266	0.166	0.192	0.05	0.241	0.201
	[0.606]	[0.606]	[0.684]	[0.662]	[0.82]	[0.624]	[0.654]
$Q(5)$	4.920	4.920	4.43	4.60	1.46	4.87	4.67
	[0.426]	[0.426]	[0.489]	[0.466]	[0.92]	[0.432]	[0.457]
$Q(10)$	13.30	13.30	13.00	13.20	8.41	13.40	13.30
	[0.206]	[0.206]	[0.222]	[0.589]	[0.80]	[0.200]	[0.207]
$Q(20)$	20.60	20.60	20.30	20.30	15.50	20.60	20.40
	[0.423]	[0.423]	[0.451]	[0.439]	[0.744]	[0.421]	[0.432]

Note: Standard errors are shown in parentheses and the *p*-values in brackets.

4.5. INCOME

from the LC may have bounded support. Since the distribution's tails significantly impact the estimates of inequality and poverty, this might be a significant issue. An additional limitation is that some families of LCs are not able to characterize the whole distribution of income, but specific parts (see for example Sarabia et al., 2010b). For example, the LC family developed by Basmann et al. (1990) fits the lower tail of the distribution accurately but gives a poor approximation at the tail's top.

On the other hand, Leimkuhler's curve evaluates the inequality level but looks at the proportion of income accrued at the top of the distribution. A Leimkuhler curve is a powerful tool for analyzing data in informetrics (see, for instance, Burrell, 2005). Both plot the cumulative proportion of total productivity against the cumulative proportion of sources. Hence, Lorenz and Leimkuhler's curves are closely related, with the only difference being that the contributions are sorted in decreasing order in Leimkuhler's curve. For instance, if the Lorenz curve suggested that 40% of the contributions produced 20% of the publication credit, the Leimkuhler curve would report that 40% of the contributions produced 80% of the publication credit.

4.5.1 Basic elements

Let $\mathcal{L}$ be the class of all non-negative random variables with positive finite expectations. For a random variable X in $\mathcal{L}$ with cdf F_X, we define its inverse distribution function by

$$F_X^{-1}(x) = \sup\{x : F_X(x) \leq x\}, \quad x \in [0,1].$$

The mathematical expectation of X is $\mu_X = \int_0^1 F_X^{-1}(y)\,dy$. The LC of X can be described as the set of points,

$$(F(x), F_{(1)}(x)) \tag{4.24}$$

defined in the unit square, where x ranges from 0 to ∞ completed if necessary by linear interpolation. For each positive x, $F(x)$ represents or approximates the proportion of individuals in the population whose income is less than or equal to x. $F_{(1)}(x)$, defined by

$$F_{(1)}(x) = \frac{\int_0^x y\,dF(y)}{\int_0^\infty y\,dF(y)}$$

is the first-moment distribution of the population, which represents the proportion of the total incomes that accrues to individuals with incomes less than or equal to x.

An expression for the LC can be constructed using the parametric representation in (4.24):

$$L(p) = F_{(1)}(F^{-1}(p)).$$

In advance, we will work with the next definition of LC given by Gastwirth (1971).

Definition 4.1 Let $X \in \mathcal{L}$ with cdf F_X and inverse distribution function F_X^{-1}. The LC of X, L_X, is defined by

$$L_X(p) = \frac{1}{\mu_X} \int_0^p F_X^{-1}(y)\,dy, \quad 0 \leq p \leq 1. \tag{4.25}$$

This definition can be applied to a finite population and to continuous variables. In equation (4.25), F_X^{-1} is piecewise continuous, and the integral can be assumed to be an ordinary Riemann integral. From definition (4.25), we can show that the LC satisfies the following properties that characterize a genuine LC:

(i) $L_X(p)$ is a continuous function on p.

(ii) $L_X(p)$ is a non-decreasing function on p.

(iii) $L_X(p)$ is a convex function.

(iv) $L_X(p)$ is differentiable almost everywhere in $[0,1]$.

(v) $L_X(p)$ verifies that $L(0) = 0$ and $L(1) = 1$.

The following theorem provides a formal definition of these properties.

Theorem 4.1 Suppose $L(p)$ is defined and continuous on $[0,1]$ with second derivative $L''(p)$. The function $L(p)$ is a LC if and only if

$$L(0) = 0, \quad L(1) = 1, \quad L'(0^+) \geq 0, \quad L''(p) \geq 0 \text{ in } (0,1). \tag{4.26}$$

4.5.2 Inequality measures and population functions

The Gini coefficient (also known as the Lorenz concentration ratio) is a measure (degree of concentration) of the inequality of a variable when we consider the distribution of its elements, on a scale from 0 to 1. It is defined, see for instance Kleiber and Kotz (2003), as

$$G = 1 - 2\int_0^1 L(p)\,dp,$$

An important generalization of the Gini index was proposed by Yitzhaki (1983), who suggested the generalized Gini index, which is defined as

$$G_\nu = 1 - \nu(\nu - 1)\int_0^1 (1-p)^{\nu-2} L(p)\,dp, \tag{4.27}$$

where $\nu > 1$ and $L(p)$ is the Lorenz curve. Of course, if $\nu = 2$ we obtain the Gini index.

An exciting but less well-known index of inequality is the Pietra index, given by the proportion of total income that would need to be reallocated across the population to achieve perfect income equality. This proportion is given by

$$P = \max_{0 \leq p \leq 1}\,[p - L(p)] = \frac{1}{2\mu} E|X - \mu|$$

and corresponds to the maximal vertical deviation between the Lorenz curve and the egalitarian line (Pietra , 1915). The value of p from which we get P can be obtained by solving the equation $\frac{d}{dp}(p - L(p)) = 0$.

The LC is related to the distribution of X by the following equation

$$F_X^{-1}(x) = \mu_X L'(x), \tag{4.28}$$

which can be obtained from (4.25) by deriving on both sides of the equality and applying the Fundamental Theorem of Calculus. From (4.28) we can get easily the cdf, F_X. The pdf $f_X(x)$ is also associated with the LC $L(p)$ (see, for example, Arnold, 1987). If $L''(p)$ exists and is positive everywhere in an interval (x_1, x_2), then F_X has a finite positive density in the interval $(\mu L'(x_1^+), \mu L'(x_2^-))$ which is given by

$$f_X(x) = \frac{1}{\mu L''(F_X(x))}. \tag{4.29}$$

4.5.3 Lorenz ordering

For the comparison of estimated income distributions, it is of interest to know the parameter values for which LCs do or do not intersect. Lorenz ordering is an essential aspect of analyzing income and wealth distributions. If we define L to be the class of all non-negative random variables with positive finite expectation, the Lorenz partial order $\leq_L$ on the class L is defined by

$$X \leq_L Y \iff L_X(p) \geq L_Y(p), \forall p \in [0, 1].$$

If $X \leq_L Y$, then X exhibits less inequality than Y in the Lorenz sense. Other stronger definitions of stochastic orderings are given in the statistical literature related with LC (see for instance, Sarabia, 2008b). The ordering defined above is a partial order invariant concerning scale transformation.

Example 4.1 *Assume that the income X follows a classical Pareto distribution with survival function given by $\bar{F}_X(x) = (\sigma/x)^\alpha$, $x \geq \sigma > 0$, $\alpha > 0$. It is simple to see that $F_X^{-1}(y) = \sigma(1-y)^{-1/\alpha}$. Since $\mu_X = \alpha\sigma/(\alpha - 1)$, $\alpha > 1$, using (4.25) we get*

$$L_\alpha(p) = \frac{\alpha - 1}{\alpha\sigma} \int_0^p \sigma(1-y)^{-1/\alpha} \, dy = 1 - (1-p)^{1-1/\alpha}, \quad 0 \leq p \leq 1. \tag{4.30}$$

Some representations of the LC in (4.30) are shown in Figure 4.3 for special values of the parameter $\alpha > 1$.

Using (4.27) we get the Yitzhaki index for the Pareto LC which results $G_\nu = (\nu - 1)/(\nu\alpha - 1))$. The corresponding Gini index is $G = 1/(2\alpha - 1)$. The value of p for which the maximum of $[p - L(p)]$ is achieved is $1 - \left(\frac{\alpha-1}{\alpha}\right)^\alpha$. Thus, some computations provide the Pietra index which results

$$P = \frac{1}{\alpha - 1}\left(\frac{\alpha - 1}{\alpha}\right)^\alpha.$$

On the other hand, it is a simple exercise to see that if $\alpha_1 \geq \alpha_2$ then $L_{\alpha_1}(p) \leq L_{\alpha_2}(p)$, i.e., the LC based on the classical Pareto distribution is ordered with respect to the parameter α.

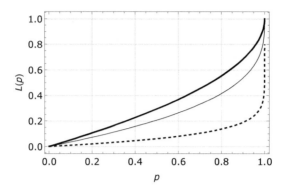

Figure 4.3: Pareto LC for special values of its parameters: $\alpha = 1.1$ (dashed), $\alpha = 1.5$ (thin) and $\alpha = 2$ (thick).

4.5.4 Estimation

Parameters of the LC can be performed by non-linear least squares, minimizing the sum of the squared differences between the theoretical models and the empirical points of the LC. For a given LC, $L_\Theta(p)$, the minimization problem can be expressed as

$$\min_{\Theta} \sum_{i=1}^{n} (p_i - L_\Theta(p_i))^2, \tag{4.31}$$

where the points $(p_i, L_\Theta(p_i))_{i=1}^n$ are the empirical points of the LC and Θ is the vector of parameters to be estimated. For comparisons of different LC fitted to a set of empirical data we can use the following measures,

$$\text{MAX} = \max_{i=1,\ldots,n} \left| q_i - L(p_i, \widehat{\Theta}) \right|,$$

$$\text{MAE} = \frac{1}{n} \sum_{i=1}^{n} \left| q_i - L(p_i, \widehat{\Theta}) \right|,$$

denoting the maximum absolute error and mean absolute error, respectively. Additionally, the sum of squares of model estimation errors (SSE) is a good measure of comparison.

From (4.28) or (4.29), the maximum likelihood estimation based on the use of the population function can also be studied. For example, when data are grouped, let n_i be the number of observations in the interval $(c_{j-1}, c_j]$. The log-likelihood function is then,

$$\ell(\Theta) = \sum_{i=1}^{n} n_i \log \left[F(x_i|\Theta) - F(x_{i-1}|\Theta) \right],$$

where n is the sample size and Θ the parameter/s to be estimated. See Chotikapanich (2008) for details. Because there is a mapping from the LC to the density of the

4.5. INCOME

data and correcting standard errors for model misspecification, the parameters of interest can be estimated by maximizing the log-likelihood to get robust (sandwich) standard errors. See Freedman (2006) for details.

Finally, when the population function associated with a given Lorenz curve is not known, estimation based on the use of the Dirichlet distribution is adequate for comparing different models (see Chotikapanich and Griffiths, 2002).

The development of new functional forms of Lorenz curves has been an attractive area of research in recent decades; see, for example, Kakwani (1980), Gupta (1984), McDonald (1984), Basmann et al. (1990), Chotikapanich (1993), Sarabia et al. (1999) and Sarabia et al. (2010a). For a recent review of Lorenz curves and income distributions, see also Chotikapanich (2008) and Arnold and Sarabia (2018). These methods also provide new functional forms of Leimkuhler curves, which are interesting in terms of informetrics and, in particular, regarding concentration aspects in this field (see Burrell, 1992, Burrell, 2005, Sarabia and Sarabia, 2008, Sarabia et al., 2010a, Gómez-Déniz, 2016 and Gómez-Déniz et al., 2021, among others). Some classical parametric LCs are shown in Table 4.11.

Table 4.11: Some classical parametric LCs.

Name	LC
Egalitarian	$L(p) = p$
Power	$L(p) = p^\alpha$, $\alpha \geq 1$
Gupta	$L(p) = p \exp[-\alpha(1-p)]$, $\alpha \geq 0$
Kakwani & Podder	$L(p) = p^\alpha \exp[-\beta(1-p)]$, $\alpha \geq 1$, $\beta > 0$
Chotikapanich	$L(p) = \frac{\exp(\alpha p) - 1}{\exp(\alpha) - 1}$, $\alpha > 0$

Example 4.2 *In this example, we compare the performance of the Pareto, power and Gupta (see Table 4.11) LCs by using real income data. We use US data on individual incomes in 1977 which was retrieved from Ryu and Slottje (1996), which provides grouped data in the form of income shares. This set of data is shown in Table 4.12.*

Table 4.12: Data for cross-sectional family, in the U.S.A. (source Ryu and Slottje, 1996).

Population, p	Income	Population, p	Income
0.10	0.0122	0.60	0.2870
0.20	0.0379	0.70	0.3956
0.30	0.0770	0.80	0.5311
0.40	0.1303	0.90	0.7062
0.50	0.1996	0.99	0.9487

The estimation is performed by non-linear least squares, minimizing the sum of the squared differences between the theoretical models and the empirical points of the

Table 4.13: Results for the parameters estimates and MSE and MAX criteria based on 1977 CPS data for cross-sectional family, in the U.S.A. (source Ryu and Slottje, 1996).

LC	$\hat{\alpha}$	SE	SSE	MAX	MAE	Gini	Pietra
Pareto	2.019	0.095	0.0590	0.0543	0.0518	0.3289	0.2466
Power	2.783	0.129	0.0275	0.0543	0.0349	0.4713	0.3608
Gupta	2.145	0.091	0.0106	0.0326	0.0223	0.4514	0.3457

LC according to (4.31). The result appears in Table 4.13 in which are also shown several error measures, the SSE, the maximum absolute error (MAX), and the mean absolute error (MAE). It can be seen that the Gupta LC provides better results in terms of smaller SSE, MAX, and MAE. Furthermore, the Gini and Pietra are the closest to the empirical ones. The empirical Gini was computed according to Brown's formula, given by $\tilde{G} = 1 - \sum_{i=1}^{n-1}(p_{i+1} - p_i)(q_{i+1} + q_i)$ and the empirical Pietra index according to $\tilde{P} = \max_{1 \leq i \leq n}(p_i - q_i)$. In this case, they result in 0.4323 and 0.313, respectively, for the data considered. Gini and Pietra indices associated with the Gupta LC is better in comparison to the ones obtained with the Pareto and power LCs.

We also present a graphical illustration of the accuracy of each model in Figure 4.4. As expected, Gupta LC fit reasonably better than the empirical LC in comparison with the Pareto and power LCs.

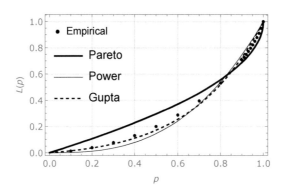

Figure 4.4: Empirical and fitted Lorenz curves based on 1977 CPS data for cross-sectional family, in the U.S.A.

Example 4.3 *Consider the function*

$$L(p) = \frac{p}{1 + \alpha(1-p)}, \quad 0 \leq p \leq 1, \ \alpha > 0. \tag{4.32}$$

In order to see that $L(p)$ represents a genuine LC we check the conditions given in (4.26):

- $L(0) = 0$ and $L(1) = 1$.

4.5. INCOME

- $L'(p) = (1+\alpha)/[1+\alpha(1-p)]^2 > 0$.
- $L''(p) = 2\alpha(1+\alpha)/[1+\alpha(1-p)]^3 > 0$.

To compute the population distribution we use (4.28) resulting that

$$F^{-1}(p) = \frac{\mu(1+\alpha)}{[1+\alpha(1-p)]^2},$$

from which it is a simple exercise to see that

$$F(x) = 1 + \frac{1}{\alpha}\left[1 - \sqrt{\frac{\mu(1+\alpha)}{x}}\right], \quad \frac{\mu}{1+\alpha} \leq x \leq \mu(1+\alpha),$$

where μ is the mean of the population distribution and the support of x is obtained by imposing that $0 \leq F(x) \leq 1$.

Expression given in (4.32) corresponds to the proposal of Aggarwal (1984) and representations of the same are provided in Figure 4.5 for special values of the parameter $\alpha > 0$.

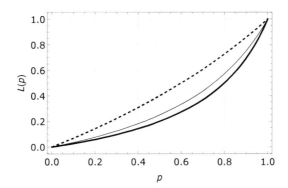

Figure 4.5: Aggarwal LC for special values of its parameters. $\alpha = 0.5$ (dashed), $\alpha = 2$ (thin) and $\alpha = 3$ (thick).

4.5.5 Leimkuhler curve

A curve related with the LC is the Leimkuhler curve. Given a LC, say $L(p)$, the relationship between the Lorenz and the Leimkuhler curves is determined by (see for example, Sarabia and Sarabia, 2008, Sarabia et al., 2010a, Sarabia, 2008a and Balakrishnan et al., 2010, among others)

$$K_0(p) = 1 - L(1-p), \quad 0 \leq p \leq 1. \tag{4.33}$$

Next definition proposed by Sarabia and Sarabia (2008) makes use of the inverse of the cdf, $F(x)$ of a random variable X.

Definition 4.2 Let X be a random variable with cdf $F_X(x)$ and inverse distribution function $F_X^{-1}(x)$. The Leimkuhler curve K_X corresponding to X is defined by

$$K_X(u) = \frac{1}{\mu_X} \int_{1-u}^{u} F_X^{-1}(y)\, dy, \qquad (4.34)$$

where μ_X is the mathematical expectation of X.

From definition (4.34) it can be easily seen that a Leimkuhler curve will be a continuous, non-decreasing concave function that is differentiable almost everywhere in $[0,1]$ with $K_X(0) = 0$ and $K_X(1) = 1$.

Example 4.4 Let the LC provided in (4.30). It is a simple exercise to see that by using (4.33) or (4.34) the Leimkuhler function obtained from this distribution is given by $L(u) = u^{1-1/\alpha}$, $\alpha > 1$. Some plots of this Leimkuhler curve are shown in Figure 4.6.

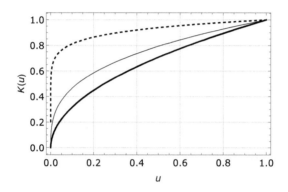

Figure 4.6: Plot of the Leimkuhler curves of the classical Pareto distribution for special values of its parameters. $\alpha = 1.1$ (dashed), $\alpha = 1.5$ (thin) and $\alpha = 2$ (thick).

Exercises

1. Consider the stochastic cost frontier model with normal and half normal distributions, i.e., $\nu = -u + \varepsilon$. Show that the marginal and conditional distributions are given by,

$$f_{\sigma_u, \sigma_\nu}(\varepsilon) = \frac{2}{\sigma} \phi\left(\frac{\varepsilon}{\sigma}\right) \Phi\left(\frac{\varepsilon \lambda}{\sigma}\right),$$

$$f(u|\varepsilon) = \frac{1}{\sqrt{2\pi}\sigma_*} \left[\Phi\left(-\frac{\mu_*}{\sigma_*}\right)\right]^{-1} \exp\left\{-\frac{(u+\mu_*)^2}{2\sigma_*^2}\right\}.$$

4.5. INCOME

2. Determine that the mean and variance of the marginal $f(\varepsilon)$ for the model provided in the previous problem are given by

$$E(\varepsilon) = \sigma_u \sqrt{\frac{2}{\pi}},$$

$$var(\varepsilon) = \sigma_\nu^2 + \sigma_u^2 \left(1 - \frac{2}{\pi}\right),$$

respectively.

3. Consider the stochastic production frontier model with normal and gamma distributions, i.e., $\nu = u + \varepsilon$ with $f(u) = u^{\lambda-1} \exp(-u/\sigma_u)/(\sigma_u^\lambda \Gamma(\lambda))$, $\lambda > 0$, $\sigma_u > 0$. Show that the marginal distribution of ε is given by,

$$f(\varepsilon) = \frac{\sqrt{2\pi}\phi(\varepsilon/\sigma_\nu)}{\sqrt{2^\alpha}(\sigma_u/\sigma_\nu)^\alpha \sigma_\nu} \sum_{j=1}^{2} R_j(\alpha, \sigma_u, \sigma_\nu, \varepsilon),$$

where

$$R_j(\alpha, \sigma_u, \sigma_\nu, \varepsilon) = (-1)^{j-1} {}_1F_1\left(\frac{\alpha+j-1}{2}; j-\frac{1}{2}; z(\sigma_u, \sigma_\nu, \varepsilon)\right),$$

$$z(\sigma_u, \sigma_\nu, \varepsilon) = \frac{\sigma_\nu^2}{2}\left(\frac{1}{\sigma_u} + \frac{\varepsilon}{\sigma_\nu^2}\right)^2$$

and ${}_1F_1$ is the Kummer confluent hypergeometric function given in (1.19).

4. Show that if X follows a Lindley distribution with pdf

$$f(x) = \frac{\theta^2}{1+\theta}(1+x)\exp(-\theta x), \quad x > 0,\ \theta > 0,$$

the log-likelihood function of the Lindley ACD(1,1) model is given by,

$$\ell(\tilde{x}; \Theta) = \log\theta + \log(\theta+2) - 2\log(1+\theta) - \log\psi_i$$
$$+ \log\left(1 + \frac{x_i}{\psi_i\,\phi(\theta)}\right) - \frac{\theta\, x_i}{\psi_i\,\phi(\theta)},$$

where $\phi(\theta) = \theta(\theta+1)/(\theta+2)$.

5. Show that if X follows an exponential distribution with mean $\lambda > 0$ and λ has an inverse Gaussian distribution with pdf given by (4.21) then the unconditional distribution of X has pdf given by (4.23) with mean $\mu = \gamma$ and variance $\sigma^2 = \gamma^2(1 + 2\gamma/\theta)$.

6. Show that the Yitzhaki index for the Gupta LC is given by $G_\nu = 1 - \alpha\nu B(\alpha, \nu)$.

7. Let $L(p)$ a genuine LC and consider the function $H(p) = \frac{d}{dp}L(p)$. Prove that $\int_0^p H(p)\,dp = 1$.

8. Consider the Power LC given by $L(p;\alpha) = p^\alpha$, with $\alpha > 1$ for which this parameter is random (see for instance, Sarabia et al., 2005) following the pdf given by

$$g(\alpha) = \frac{\lambda^\sigma}{\Gamma(\sigma)}(\alpha-1)^{\sigma-1}\exp[-\lambda(\alpha-1)], \quad \alpha > 1,$$

i.e., a translated gamma distribution with parameters $\lambda > 0$ and $\sigma > 0$. Show that the mixture obtaining by $\int_1^\infty L(p;\alpha)g(\alpha)\,d\alpha$ is a genuine LC with expression

$$L(p) = p\left(1 - \frac{1}{\lambda}\log p\right)^{-\alpha}$$

and with Gini index given by $G = 1 - 2\lambda\exp(2\lambda)E_\sigma(2\lambda)$, where $E_n(z) = \int_1^\infty \tau^{-n}\exp(-z\tau)\,d\tau$ is the exponential integral function.

9. Prove that the population sizes of the city corresponding to the 80-th percentile in French cities for the years 1962, 1975, 1990, 1999, 2006 and 2012 using the composite lognormal-Pareto model are:

	1962	1975	1990	1999	2006	2012
80th Percentile	162	141	146	149	163	170

10. Given the Aggarwal LC in (4.32), show the following:

    (a) If $\alpha_1 \geq \alpha_2$ then $L_{\alpha_1}(p) \leq L_{\alpha_2}(p)$, i.e., the Aggarwal LC is ordered with respect to the parameter α.

    (b) Use the data given in Table 4.12 to show that the fit of the Aggarwal LC to the same provided the following result,

LC	$\hat{\alpha}$	SE	SSE	MAX	MAE	Gini	Pietra
Aggarwal	2.8645	0.081	0.0590	0.0543	0.0518	0.3289	0.2466

Bibliography

Abbring, J. and van den Berg, G. 2007. The unobserved heterogeneity distribution in duration analysis. *Biometrika*, 94(1): 87–99.

Abramowitz, M. and Stegun, I. 1972. *Handbook of Mathematical Functions*. New York: Dover Publications, Inc.

Acerbi, C. and Tasche, D. 2002. On the coherence of expected shortfall. *Journal of Banking & Finance*, 26: 1487–1503.

Aggarwal, V. 1984. On optimum aggregation of income distribution data. *Sankhyā: The Indian Journal of Statistics, Series B*, 46: 343–355.

Aggarwal, V. and Singh, R. 1984. On optimum stratification with proportional allocation for a class of Pareto distributions. *Sankhyā: The Indian Journal of Statistics, Series B*, 13: 3017–3116.

Agarwal, V. and Yochum, G. 1999. Tourist spending and the race of visitors. *Journal of Travel Research*, 38(2): 173–176.

Aguiló, E. and Juaneda, C. 2000. Tourist expenditure for mass tourism markets. *Annals of Tourism Research*, 27(3): 624–637.

Aguiló, E., Rosselló, J. and Vila, M. 2017. Length of stay and daily tourist expenditure: A joint analysis. *Tourism Management Perspectives*, 21: 10–17.

Aigner, D., Lovell, C. and Schmidt, P. 1977. Formulation and estimation of stochastic frontier production function model. *Journal of Econometrics*, 12: 21–37.

Albrecher, H., Beirlant, J. and Teugels, J.L. 2017. *Reinsurance: Actuarial and Statistical Aspects*. Wiley Series in Probability and Statistics. Wiley. Hoboken.

Alegre, J., Mateo, S. and Pou, L. 2006. The length of stay in the demand for tourism. *Tourism Management*, 27: 1343–1355.

Alegre, J., Mateo, S. and Pou, L. 2011. A latent class approach to tourists' length of stay. *Tourism Management*, 32: 555–563.

Alegre, J. and Pou, L. 2003. The reduction of the length of stay in holiday destinations: Implications on tourist expenditure and seasonality in the Balearic Islands (in Spanish). In *Situación, Serie Estudios Regionales*, pp. 177–202. BBVA.

Alegre, J. and Pou, L. 2007. Microeconomic determinants of the duration of stay of tourists. pp. 181–206. *In*: Matias, A., Nijkamp, P. and Neto, P. (eds.). *Advances in Modern Tourism Research*. Heidelberg: Physica-Verlag.

Alegre, J. and Garau, J. 2011. The factor structure of tourist satisfaction at sun and sand destinations. *Journal of Travel Research*, 50(1): 78–86.

Anderson, G. and Ge, Y. 2005. The size distribution of Chinese cities. *Regional Science and Urban Economics*, 35(6): 756–776.

Arnold, B.C. 1983. Pareto Distributions. *International Cooperative Publishing House, Silver Spring, MD*.

Arnold, B.C. 1987. *Majorization and the Lorenz Curve: A Brief Introduction, Lecture Notes in Statistics, 43*. Springer-Verlag, Berlin.

Arnold, B.C. and Sarabia, J.M. 2018. *Majorization and the Lorenz Order with Applications in Applied Mathematics and Economics*. Springer.

Azzalini, A. 1985. A class of distributions which includes the normal ones. *Scandinavian Journal of Statistics*, 12: 171–178.

Azzalini, A. 2013. *The Skew-normal and Related Families*. Cambridge University Press, Cambridge.

Azzalini, A., Cappello, D. and Kotz, S. 2003. Log-skew-normal and log-skew-t distributions as model for family income data. *Journal of Income Distribution*, 11: 12–20.

Balakrishnan, N. and Nevzorov, V.B. 2003. A Primer on Statistical Distributions. *Wiley and Sons, New York*.

Balakrishnan, N., Sarabia, J. and Kolev, N. 2010. A simple relation between the Leimkuhler curve and the mean residual life. *Journal of Informetrics*, 4(4): 602–607.

Basmann, R.L., Hayes, K.L., Slottje, D.J. and Johnson, J.D. 1990. A general functional form for approximating the Lorenz curve. *Journal of Econometrics*, 43: 77–90.

Battese, G. and Coelli, T. 1988. Prediction of firm-level technical efficiencies with a generalized frontier production function and panel data. *Journal of Econometrics*, 38: 387–399.

Battese, G. and Coelli, T. 1995. A model for technical inefficiency effects in a stochastic frontier production function for panel data. *Empirical Economics*, 20: 325–332.

Battese, G. and Corra, G. (1977). Estimation of a production frontier model: With application to the pastoral zone of Eastern Australia. *Australian Journal of Agricultural Economics*, 21(3): 169–179.

Bhattacharya, S.K. and Kumar, S. 1986. E-IG model in life testing. *Calcutta Statistical Association Bulletin*, 35: 85–90.

Bhatti, C.R. 2009. On the interday homogeneity in the intraday rate of trading. *Mathematics and Computers in Simulation*, 79: 2250–2257.

Bhatti, C. 2010. The Birnbaum-Saunders autoregressive conditional duration model. *Mathematics and Computers in Simulation*, 80: 2036–2078.

Bose, S., Shmueli, G., Sur, P. and Dubey, P. 2013. Fitting Com-Poisson mixtures to bimodal count data. *Proceedings of the 2013 International Conference on Information, Operations Management and Statistics (ICIOMS2013)*, pp. 1–8.

Bozdogan, H. 1987. The general theory and its analytical extension. *Psychometrika*, 52(3): 345–370.

Brida, J., Disegna, M. and Osti, L. 2013. Visitors' expenditure behaviour at cultural events: The case of Christmas markets. *Tourism Economics*, 19(5): 1173–1196.

Brida, J. and Scuderi, R. 2012. Determinants of tourist expenditure: A review of micro econometric models. *Munich Personal RePEc Archive (MPRA)* (Paper 38468).

Brooks, C. 2009. *RATS Handbook to Accompany Introductory Econometrics for Finance*. Cambridge University Press.

Bull, A. 1995. *The Economics of Travel and Tourism (2nd ed.)*. Australia: Longman.

Burrell, Q.L. 1991. The Bradford distribution and the Gini index. *Scientometrics*, 21: 181–194.

Burrell, Q.L. 1992. The Gini index and the Leimkuhler curve for bibliometric processes. *Information Processing and Management*, 28: 19–33.

Burrell, Q.L. 2005. Summetry and other transformation features of Lorez/Leimkuhler representations of informetric data. *Information Processing and Management*, 41: 1317–1329.

Burrell, Q.L. 2006. On Egghe's version of continuous concentration theory. *Journal of the American Society for Information Science and Technology*, 57: 1406–1411.

Calderín-Ojeda, E. 2015. On the composite Weibull–Burr model to describe claim data. *Communications in Statistics: Case Studies, Data Analysis and Applications*, 1(1): 59–69.

Calderín-Ojeda, E. 2016. The distribution of all French communes: A composite parametric approach. *Physica A*, 450: 385–394.

Calderín-Ojeda, E., Azpitarte, F. and Gómez-Déniz, E. 2016. Modelling income data using two extensions of the exponential distribution. *Physica A*, 461: 756–766.

Calderín-Ojeda, E. and Kwok, C.F. 2016. Modeling claims data with composite Stoppa models. *Scandinavian Actuarial Journal*, 2016(9): 817–836.

Cambanis, S. 1977. Some properties and generalizations of multivariate Eyraud–Gumbel–Morgenstern distributions. *Journal of Multivariate Analysis*, 7: 551–559.

Cannon, T. and Ford, J. 2002. Relationship of demographic and trip characteristics to visitor spending: an analysis of sports travel visitors across time. *Tourism Economics*, 8(3): 263–271.

Cárdenas, P., Pulido, J. and Pulido, M. 2015. The influence of tourist satisfaction on tourism expenditre in emerging urban cultural destinations. *Journal of Travel & Tourism Marketing*, 33(497): 497–512.

Chen, J. and Novick, M. 1984. Bayesian analysis for binomial models with generalized beta prior distributions. *Journal of Educational Statistics*, 9: 163–175.

Chhikara, R.S. and Folks, J.L. 1977. The inverse Gaussian distribution as a life time model. *Technometrics*, 19: 461–468.

Chhikara, R.S. and Folks, J.L. 1989. The inverse Gaussian Distribution. Theory, Methodology, and Applications. *Marcel Dekker, New York*.

Cho, J. and White, H. 2010. Testing for unobserved heterogeneity in exponential and Weibull duration models. *Journal of Econometrics*, 157(2): 458–480.

Chotikapanich, D. 1993. A comparison of alternative functional forms for the Lorenz curve. *Economics Letters*, 41: 129–138.

Chotikapanich, D. 2008. *Modeling Income Distributions and Lorenz Curves*. Springer.

Chotikapanich, D. and Griffiths, E. 2002. Estimating Lorenz curves using a Dirichlet distribution. *Journal of Business & Economic Statistics*, 20(2): 290–295.

Coelli, T., Rao, D.S.P. and Battese, G.E. 2003. *An Introduction to Efficiency and Productivity Analysis*. Kluwer Academic Publishers.

BIBLIOGRAPHY

Coelli, T.J., Prasada Rao, D.S., O'Donnell, C. and Battese, G.E. 2005. *An Introduction to Efficiency and Productivity Analysis*. Springer-Verlag, Berlin, 2nd Edition.

Consul, P.C. 1974. A simple urn model dependent upon predetermined strategy. *Sankhyā: The Indian Journal of Statistics, Series B*, 36(4): 391–399.

Consul, P.C. 1990. On some properties and applications of quasi-binomial distribution. *Communications in Statistics-Theory and Methods*, 19(2): 477–504.

Consul, P. and Mittal, S. 1973. A generalization of the Poisson distribution. *Technometrics*, 15(4): 791–799.

Cooray, K. and Ananda, M. 2005. Modeling actuarial data with a composite lognormal-Pareto model. *Scandinavian Actuarial Journal*, 2005(5): 321–334.

Craggs, R. and Schofield, P. 2009. Expenditure-based segmentation and visitor profiling at the Quays in Salford, uk. *Tourism Economics*, 15(1): 243–260.

Decrop, A. and Snelders, D. 2004. Planning the summer vacation an adaptable process. *Annals of Tourism Research*, 31(4): 1008–1030.

Dhaene, J., Denuit, D., Goovaerts, M.J., Kaas, R. and Vyncke, D. 2002. The concept of comonotonicity in actuarial science and finance: Theory. *Insurance: Mathematics and Economics*, 31(1): 3–33.

Dhaene, J., Vanduffel, S., Goovaerts, M.J., Kaas, R., Tang, Q. and Vyncke, D. 2006. Risk measures and comonotonicity: A review. *Stochastic Models*, 22(4): 573–606.

Denuit, M., Dhaene, J., Goovaerts, M. and Kaas, R. 2005. *Actuarial Theory for Dependent Risks*. John Wiley & Sons Ltd, The Atrium, Southern Gate, Chichester, West Sussex PO19 8SQ, England.

Dickson, D.C.M. 2017. *Insurance Risk and Ruin. Second Edition*. International Series on Actuarial Science. Cambridge University Press. Cambridge.

Dean, C., Lawless, J.F. and Willmot, G.E. 1989. A mixed Poisson-Inverse Gaussian regression model. *The Canadian Journal of Statistics*, 17(2): 171–181.

Disegna, M. and Osti, L. 2016. Tourists' expenditure behaviour: The influence of satisfaction and the dependence of spending categories. *Tourism Economics*, 22(1): 5–30.

Eeckhout, J. 2004. Gibrat's Law for cities: An explanation. *American Economic Review*, 94(5): 1429–1451.

Edgell, D. and Swanson, J. 2013. *Tourism Policy and Planning. Yesterday, Today, and Tomorrow*. Taylor & Francis Ltd—Routledge: London, United Kingdom.

Egghe, L. 2005. Zipfian and Lotkaian continuous concentration theory. *Journal of the American Society for Information Science and Technology*, 56: 935–945.

Engle, R. and Russell, J. 1998. Autoregressive conditional duration: A new model for irregularly-spaced transaction data. *Econometrica*, 66: 1127–1162.

Engle, R.F. and Gallo, G.M. 2006. A multiple indicators model for volatility using intra-daily data. *Journal of Econometrics*, 131: 3–27.

Engle, R.F. and Russell, J.R. 2006. Autoregressive conditional duration: a new model for irregularly-spaced transaction data. *Econometrica*, 66: 1127–1162.

Famoye, F. 2010. On the bivariate negative binomial regression model. *Journal of Applied Statitics*, 37(6): 969–981.

Farlie, D. 1960. The performance of some correlation coefficients for a general bivariate distribution. *Biometrika*, 47: 307–323.

Ferrari, S. and Cribari-Neto, F. 2004. Beta regression for modelling rates and proportions. *Journal of Applied Statistics*, 31(7): 799–815.

Ferrer-Rosell, B., Coenders, G. and Martínez-García, E. 2016. Segmentation by tourist expenditure composition: An approach with compositional data analysis and latent classes. *Tourism Analysis*, 21(6): 589–602.

Filippini, M. and Greene, W. 2016. Persistent and transient productive inefficiency: A maximum simulated likelihood approach. *Journal of Productivity Analysis*, 45: 187–196.

Frangos, N. and Karlis, D. 2004. Modelling losses using an exponential-inverse Gaussian distribution. *Insurance: Mathematics and Economics*, 35: 53–67.

Fredman, P. 2008. Determinants of visitor expenditures in mountain tourism. *Tourism Economics*, 14(2): 297–311.

Freedman, D.A. 2006. On the so-called "Huber sandwich estimator" and "robust standard errors". *The American Statistician*, 60(4): 299–302.

Gangopadhyay, K. and Basu, B. 2009. City size distributions for India and China. *Physica A*, 388(13): 2682–2688.

García-Sánchez, A., Fernández-Rubio, E. and Collado, M. 2013. Daily expenses of foreign tourists, length of stay and activities: Evidence from Spain. *Tourism Economics*, 19(3): 613–630.

Gastwirth, J.L. 1971. A general definition of the Lorenz curve. *Econometrica*, 39: 1037–1039.

BIBLIOGRAPHY

Gerber, H. 1979. *An Introduction to Mathematical Risk Theory.* Huebner Foundation Monograph 8.

Gibrat, R. 1931. *Les inégalités économiques* Librairie du Recueil Sirey, Paris.

Giesen, K., Zimmermann, A. and Suedekum, J. 2010. The size distribution across all cities—Double Pareto lognormal strikes. *Journal of Urban Economics*, 68(2): 129–137.

Gómez–Déniz, E. 2010. Another generalization of the geometric distribution. *Test*, 19: 399–415.

Gómez-Déniz, E. 2016. A family of arctan Lorenz curves. *Empirical Economics*, 51(3): 1215–1233.

Gómez-Déniz, E. and Calderín-Ojeda, E. 2015a. On the use of the Pareto ArcTan distribution for describing city size in Australia and New Zealand. *Physica A*, 436: 821–832.

Gómez-Déniz, E. and Calderín-Ojeda, E. 2015b. Parameters estimation for a new generalized geometric distribution. *Communications in Statistics: Simulation and Computation*, 44(8): 2023–2039.

Gómez-Déniz, E. and Calderín-Ojeda, E. 2018. Properties and applications of the Poisson-reciprocal inverse Gaussian distribution. *Journal of Statistical Computation and Simulation*, 88(2): 269–289.

Gómez-Déniz, E. and Calderín-Ojeda, E. 2020. On the usefulness of the logarithmic skew normal distribution for describing claims size data. *Mathematical Problems in Engineering*, 2020: 1–9.

Gómez-Déniz, E., Gallardo, D. and Gómez, H.W. 2020. Quasi-binomial zero-inflated regression model suitable for variables with bounded support. *Journal of Applied Statistics*, 47(12): 2208–2229.

Gómez-Déniz, E. and Pérez-Rodríguez, J. 2019. Modelling distribution of aggregate expenditure on tourism. *Economic Modelling*, 78: 293–308.

Gómez-Déniz, E., Pérez-Rodríguez, J. and Boza-Chirino, J. 2020. Modelling tourist expenditure at origin and destination. *Tourism Economics*, 26(3): 437–460.

Gómez-Déniz, E. and Pérez-Rodríguez, J. 2015. Closed-form solution for a bivariate distribution in stochastic frontier models with dependent errors. *Journal of Productivity Analysis*, 43: 215–223.

Gómez-Déniz, E. and Pérez-Rodríguez, J. 2016. Conditional duration model and unobserved market heterogeneity of traders. an infinite mixture of non-exponentials. *Colombian Journal of Statistics*, 39(2): 307–323.

Gómez-Déniz, E. and Pérez-Rodríguez, J. 2017. Mixture inverse Gaussian for unobserved heterogeneity in the autoregressive conditional duration model. *Communications in Statistics. Theory and Methods*, Forthcoming.

Gómez-Déniz, E. and Pérez-Rodríguez, J.V. 2019. Modeling bimodality of tourist length of stay. *Annals of Tourism Research*, 75: 131–151.

Gómez-Déniz, E. and Sarabia, J.M. 2018. A family of generalised beta distributions: properties and applications. *Annals of Data Science*, 5(3): 401–430.

Gómez-Déniz, E., Dávila-Cárdenes, N. and Boza-Chirino, J. 2021. Modelling expenditure in tourism using the log-skew normal distribution. *Current Issues in Tourism* (to appear) https://doi.org/10.1080/13683500.2021.1960282

Gómez-Déniz, E. and Pérez-Rodríguez, J. 2021. Modelling dependence between daily tourist expenditure and length of stay. *Tourism Economics*, 27(8): 1615–1628.

Gómez-Déniz, E., Sordo, M. and Calderín-Ojeda, E. 2013. The log-Lindley distribution as an alternative to the Beta regression model with applications in insurance. *Insurance: Mathematics and Economics*, 54: 49–57.

Gómez-Déniz, E., Sarabia, J.M. and Jordá, V. 2021. Parametric Lorenz curves based on the beta system of distributions. *Communications in Statistics-Theory and Methods* (to appear).

Gómez-Déniz, E., Sarabia, J.M. and Calderín-Ojeda, E. 2008. Univariate and multivariate versions of the negative binomial-inverse Gaussian distributions with applications. *Insurance: Mathematics and Economics*, 42(1): 39–49.

Goryushkina, N., Gaifutdinova, T., Logvina, E., Redkin, A., Kudryavtsev, V. and Shol, Y. 2019. Basic principles of tourist services market segmentation. *International Journal of Economics and Business Administration*, VII(2): 139–150.

Gradshteyn, I. and Ryzhik, I. 1994. *Table of Integrals, Series, and Products*. 5th ed. Jeffrey, A., ed. Boston: Academic Press.

Greene, W. 1980a. Maximum likelihood estimation of econometric frontier functions. *Journal of Econometrics*, 13(1): 27–56.

Greene, W. 1980a. On the estimation of a flexible frontier production model. *Journal of Econometrics*, 13(1): 101–115.

Greene, W. 1990. A gamma distributed stochastic frontier model. *Journal of Econometrics*, 46(1): 141–164.

Greene, W. 2003. Maximum simulated likelihood estimation of the normal-gamma stochastic frontier function. *Journal of Productivity Analysis*, 19(2-3): 179–190.

Groeneveld, R. and Meeden, G. 1984. Measuring skewness and kurtosis. *The Statistician*, 33: 391–399.

Gumbel, E.J. 1960. Bivariate exponential distributions. *Journal of the American Statistical Association*, 52: 1313–1314.

Gupta, M.R. 1984. Functional forms for fitting the Lorenz curve. *Econometrica*, 52: 1313–1314.

Hannan, E.J. and Quinn, B.G. 1979. The determination of the order of an autoregression. *Journal of the Royal Statistical Society B*, 41: 190–195.

Hastie, T. and Tibshirani, R. 1986. Generalized additive models. *Statistical Science*, 1(3): 297–318.

Hellström, J. 2006. A bivariate count data model for household tourism demand. *Journal of Applied Econometrics*, 21: 213–226.

Honoré. 1990. Simple estimation of a duration model with unobserved heterogeneity. *Econometrica*, 58: 453–473.

Heckman, J. and Singer, B. 1984. A method for minimizing the impact of distributional assumptions in econometric models for duration data. *Econometrica*, 52: 271–320.

Hougaard, P. 1984. Life table methods for heterogeneous populations: Distributions describing the heterogeneity. *Biometrika*, 71: 75–83.

Johnson, N., Kotz, S. and Balakrishnan. 1997. *Discrete Multivariate Distributions*. John Wiley and Sons, New York.

Johnson, N., Kemp, A. and Kotz, S. 2005. *Univariate Discrete Distributions*. John Wiley, INC.

Jones, M. 2009. Kumaraswamy's distribution: A beta-type distribution with some tractability advantages. *Statistical Methodology*, 6(1): 70–81.

Jørgensen, B. 1982. Statistical Properties of the Generalized Inverse Gaussian Distribution. *Lecture Notes in Statistics*, 9. Springer–Verlag, New York.

Kaas, R., Goovaerts, M., Dhaene, J. and Denuit, M. 2008. *Modern Actuarial Risk Theory*. Second Edition. Springer Science & Business Media, Springer Verlag Berlin Heidelberg.

Klugman, S.A., Panjer, H.H. and Willmot, G.E. 2008. *Loss Models. From Data to Decisions*. Third Edition. John Wiley, New Jersey.

Kakwani, N. 1980. On a class of poverty measures. *Econometrica*, 48: 437–446.

Kakwani, N.C. and Podder, N. 1976. Efficient estimation of the Lorenz curve and associated inequality measures from grouped observations. *Econometrica*, (44): 137–148.

Kamp, U. 1998. On a class of premium principles including the Esscher principle. *Scandinavian Actuarial Journal*, 1: 75–80.

Kim, S., Prideaux, B. and Chon, K. 2010. A comparison of results of three statistical methods to understand the determinants of festival participants' expenditures. *International Journal of Hospitality Management*, 29: 297–307.

Kleiber, C. and Kotz, S. 2003. *Statistical Size Distributions in Economics and Actuarial Sciences*. John Wiley & Sons, Inc.

Kumbhakar, S. 1990. Production frontiers, panel data and time-varying technical inefficiency. *Journal of Econometrics*, 46: 201–212.

Kumbhakar, S.C., Parmeter, C.F. and Tsionas, E.G. 2013. A zero inefficiency stochastic frontier model. *Journal of Econometrics*, 172: 66–76.

Lakshminarayana, J., Pandit, S.N.N and Rao, K.S. 1999. On a bivariate poisson distribution Communications in Statistics: Theory and Methods, 28(2): 267–276.

Lancaster, T. 1972. A stochastic model for the duration of a strike. Journal of the Royal Statistical Society, Series A., 135(2): 257–271.

Lancaster, T. 1990. *The Econometric Analysis of Transition Data*. Cambridge University Press.

Lee, L.-F. 1983. A test for distributional assumptions for the stochastic frontier functions. *Journal of Econometrics*, 22(3): 245–267.

Lee, J. and Choi, M. 2019. Examining the asymmetric effect of multi-shopping tourism attributes on overall shopping destination satisfaction. *Journal of Travel Research*, 59(2): 295–314.

Levy, M. 2009. Gibrat's Law for (all) cities: Comment. *American Economic Review*, 99(4): 1672–1675.

Limburg, B.V. 1997. Overnight tourism in Amsterdam 1982–1993: A forecasting approach. *Tourism Management*, 18(7): 465–468.

Lin, G. and Stoyanov, J. 2009. The logarithmic skew-normal distribution are moment-indeterminate. *Journal of Applied Probability*, 46: 909–916.

Luca, G.D. and Gallo, G. 2004. Mixture processes for intradaily financial durations. *Studies in Nonlinear Dynamics and Econometrics*, 8(2): 1–18.

BIBLIOGRAPHY

Luca, G.D. and Gallo, G.M. 2009. Time-varying mixing weights in mixture autoregressive conditional duration models. *Econometric Review*, 28(1): 102–120.

Luca, G.D. and Zuccolotto, P. 2003. Finite and infinite mixtures for financial durations. *Metron*, 61: 431–455.

Luckstead, J. and Devadoss, S. 2014. A comparison of city size distributions for China and India from 1950 to 2010. *Economics Letters*, 124: 290–295.

Lunde, A. 1999. A generalized gamma autoregressive conditional duration model. *Discussion Paper, Aalborg Universiteit*.

Marcussen, C. 2011. Determinants of tourist spending in cross-sectional studies and at Danish destinations. *Tourism Economics*, 17: 833–855.

Marrocu, E., Paci, R. and Zara, A. 2015. Micro-economic determinants of tourist expenditure: A quantile regression approach. *Tourism Management*, 50: 13–30.

Marshall, A.W. and Olkin, I. 1997. A new method for adding a parameter to a family of distributions with application to the exponential and Weibull families. *Biometrika*, 84(3): 641–652.

Martínez-García, E. and Raya, J. 2008. Length of stay for low cost tourism. *Tourism Management*, 29: 1064–1075.

McDonald, J.B. 1984. Some generalized functions for the size distribution of income. *Econometrica*, 52: 647–664.

Meeusen, W. and Broeck, J.V.D. 1977. Efficiency estimation from Cobb-Douglas production function with composed error. *Econometric Reviews*, 18: 435–444.

Moura, N.J. and Ribeiro, M.B. 2006. Zipf law for Brazilian cities. *Physica A*, 367: 441–448.

Mules, T. 1998. Decomposition of australian tourist expenditure. *Tourism Management*, 19: 267–271.

Nadarajah, S. 2005. Exponentiated beta distributions. *Computers & Mathematics with Applications*, 49: 1029–1035.

Nicolau, J. and Más, F. 2004. A random parameter logit approach to the two-stage tourist choice process: Going on holidays and length of stay.

Nogawa, H., Yamaguchi, Y. and Hagi, Y. 1996. An empirical research study on japanese sport tourism in sport-for-all events: Case studies of a single-night event and a multiple night event. *Journal of Travel Research*, 35(4): 46–54.

O'Hara, M. 1995. *Market Microstruture Theory*. Basil Blackwell Inc., Oxford, 1995.

Pani, A., Sahu, P. and Majumdar, B. 2020. Expenditure-based segmentation of freight travel markets: Identifying the determinants of freight transport expenditure for developing marketing strategies. *Research in Transportation Business & Management*. In press https://doi.org/10.1016/j.rtbm.2020.100437.

Panjer, H.H. 1981. Recursive evaluation of family of compound distributions *Astin Bulletin*, 12(1): 22–26.

Philippou, A.N., Georghiou, C. and Philippou, G.N. 1983. A generalized geometric distribution and some of its properties. *Statistics & Probability Letters*, 1: 171–175.

Pietra, G. 1915. Delle relazioni tra gli indici di variabilit. *Atti Regio Istituto Veneto*, 74(II): 775–792.

Pyrlik, V. 2013. Autoregressive conditional duration as a model for financial market crashes prediction. *Physica A: Statistical Mechanics and its Applications*, 392(23): 6041–6051.

Rasche, R.H., Gaffney, J., Koo, A.Y.C. and Obs, N. 1980. Functional forms for estimating the Lorenz curve. *Econometrica*, 48: 1061–1062.

Rizzo, M.L. 2009. New goodness-of-fit tests for Pareto distributions *Astin Bulletin*, 39(2): 694–715.

Roeder, K. 1994. Little and B. DelHomme-Little. A graphical technique for determining the number of components in a mixture of normals. *Journal of the American Statistical Association*, 89: 487–495.

Rolski, T., Schmidli, H., Schmidt, V. and Teugel, J. 1999. *Stochastic Processes for Insurance and Finance*. John Wiley & Sons.

Rousseau, R. 2007. On Egghe's construction of Lorenz curves. *Journal of the American Society for Information Science and Technology*, 58: 1551–1552.

Ruskeepaa, H. 2009. *Mathematica Navigator. Mathematics, Statistics, and Graphics. Third Edition*. Academic Press, USA.

Ryu, H.K. and Slottje, D.J. 1996. Two flexible functional form approaches for approximating the Lorenz curve. *Journal of Econometrics*, 72: 251–274.

Salmasi, L., Celidoni, M. and Procidano, I. 2012. Length of stay: Price and income semi-elasticities at different destinations in Italy. *International Journal of Tourism Research*, 14: 515–530.

Sarabia, J.M. 2008a. A general definition of the Leimkuhler curve. *Journal of Informetrics*, 2(2): 156–163.

Sarabia, J.M. 2008b. Parametric Lorenz curves: Models and applications. *in Modelling Income Distributions and Lorenz Curves*, D. Chotikapanich, Editor, Chapter 9, 167–192. Springer.

Sarabia, J.M., Castillo, E. and Slottje, D.J. 1999. An ordered family of lorenz curves. *Journal of Econometrics*, 91: 43–60.

Sarabia, J.M., Castillo, E., Pascual, M. and Sarabia, M. 2005. Mixture Lorenz curves. *Economics Letters*, 89: 89–94.

Sarabia, J.M., Gómez-Déniz, E., Sarabia, M. and Prieto, F. 2010a. A general method for generating parametric Lorenz and Leimkuhler curves. *Journal of Informetrics*, 4: 424–539.

Sarabia, J.M., Jordá, V. and Trueba, C. 2019. The Lamé class of Lorenz curves. *Communications in Statistics-Theory and Methods*, 46(11): 5311–5326.

Sarabia, J.M., Prieto, F. and Sarabia, M. 2010b. Revisiting a functional form for the Lorenz curve. *Economics Letters*, 107: 249–252.

Sarabia, J.M. and Sarabia, M. 2008. Explicit expressions for the Leimkuhler curve in parametric families. *Information Processing and Management*, 44: 1808–1818.

Sarabia, J.M. and Calderín-Ojeda, E. 2018. Analytical expressions of risk quantities for composite models. *The Journal of Risk Model Validation*, 12(3): 75-101.

Sarabia, J.M. and Guillén, M. 2008. Joint modelling of the total amount and the number of claims by conditionals. *Insurance: Mathematics and Economics*, 43: 466–473.

Sarabia, J.M. and Prieto, F. 2009. The Pareto–positive stable distribution: A new descriptive model for city size data. *Physica A*, 388: 4179–4191.

Scollnik, D.P.M. 2007. On composite lognormal-Pareto models. *Scandinavian Actuarial Journal*, 1: 20–33.

Seaton, A. and Palmer, C. 1997. Understanding VFR tourism behaviour: The first five years of the United Kingdom tourism survey. *Tourism Management*, 1(6): 345–355.

Seshadri, V. 1983. The inverse Gaussian distribution: Some properties and characterizations. *Canadian Journal of Statistics*, 11: 131–136.

Seshadri, V. 1999. The Inverse Gaussian Distribution: A Case Study in Exponential Families. *Oxford Science Publications*.

Shaked, M. and Shanthikumar, J.G. 2007. *Stochastic Orders Springer Series in Statistics. Springer Science + Business Media, LLC. New York*.

Smith, M. 2008. Stochastic frontier models with dependent error components. *Econometric Journal*, 11: 172–192.

Spotts, D. and Mahoney, E. 1991. Segmenting visitors to a destination region based on the volume of their expenditures. *Journal of Travel Research*, 29(4): 24–31.

Stacy, E. 1962. A generalization of the gamma distribution. *The Annals of Mathematical Statistics*, 33(3): 1187–1192.

Stevenson, R. 1980. Likelihood functions for generalized stochastic frontier functions. *Journal of Econometrics*, 13: 57–66.

Sundt, B. and Vernic R. 2009. *Recursions for Convolutions and Compound Distributions with Insurance Applications.* Springer.

Sur, P., Shmueli, G., Bose, S. and Dubey, P. 2015. Modeling bimodal discrete data using Conway-Maxwell-Poisson mixture models. *Journal of Business & Economic Statistics*, 33(3): 352–365.

Svensson, B., Moreno, P. and Martín, D. 2011. Understanding travel expenditure by means of market segmentation. *The Service Industries Journal*, 31(10): 1683–1698.

Taylor, D., Fletcher, R. and Clabaugh, T. 1993. A comparison of characteristics, regional expenditures, and economic impact of visitors to historical sites with other recreational visitors. *Journal of Travel Research*, 32(1): 30–35.

Thrane, C. 2012. Analyzing tourists' length of stay at destinations with survival models: A constructive critique based on a case study. *Tourism Management*, 33: 126–132.

Thrane, C. 2014. Modelling micro-level tourism expenditure: Recommendations on the choice of independent variables, functional form and estimation technique. *Tourism Economics*, 20(1): 51–60.

Thrane, C. 2015. Examining tourists' long-distance transportation mode choices using a multinomial logit regression model. *Tourism Management Perspectives*, 15: 115–121.

Thrane, C. 2011. Domestic tourism expenditures: The non-linear effects of length of stay and travel party size. *Tourism Management*, 32: 46–52.

Thrane, C. and Farstad, E. 2011. Domestic tourism expenditures: The non-linear effects of length of stay and travel party size. *Tourism Management*, 32: 46–52.

Tripathi, R.C., Gupta, R.C. and White, T.J. 1987. Some generalizations of the geometric distribution. *Sankhya, Series B*, 49(3): 218–223.

BIBLIOGRAPHY

Tsay, R.S. 2002. Analysis of Financial Time Series. Financial Econometrics. *John Wiley & Sons. USA.*

Tweedie, M.C.K. 1957. Statistical properties of inverse Gaussian distributions I. *Annals of Mathematical Statistics*, 28(2): 362–377.

Vinnciombe, T. and Sou, P. 2014. Market segmentation by expenditure: An under-utilized methodology in tourism research. *Tourism Review*, 69(2): 122–136.

Vuong, Q. 1989. Likelihood ratio tests for model selection and non-nested hypotheses. *Econometrica*, 57: 307–333.

Wang, D. and Davidson, M. 2010. A review of micro-analyses of tourist expenditure. *Current Issues in Tourism*, 8(5): 333–346.

Wang, E., Little, B. and DelHomme-Little, B. 2012. Factors contributing to tourists' length of stay in Dalian northeastern China-A survival model analysis. *Tourism Management Perspectives*, 4: 67–72.

Wang, D. and Davidson, M. 2010. A review of micro-analyses of tourist expenditure. *Current Issues in Tourism*, 13(5): 507–524.

Wani, J.K. 1978. Measuring diversity in biological populations with logarithmic abundance distributions. *The Canadian Journal of Statistics*, 6(2): 219–228.

Wedel, M., Desarbo, W.S., Bult, J.R. and Ramaswamy, V.J. 1993. A latent class Poisson regression model for heterogeneous count data. *Journal of Applied Econometrics*, 8(4): 397–411.

Willmot, G.E. 1987. The Poisson-inverse Gaussian distribution as an alternative to the negative binomial. *Scandinavian Actuarial Journal*, 113–127.

Wood, S. 2017. *Generalized Additive Models. An Introduction with R. Second Edition.* CRC Press.

Wu, L., Zhang, J. and Fujiwara, A. 2013. Tourism participation and expenditure behaviour: analysis using a scobit based discrete-continuous choice model. *Annals of Tourism Research*, 40: 1–17.

Yee. T. 2015. *Vector Generalized Linear and Additive Models. With an Implementation in R.* Springer.

Yitzhaki, S. 1983. On an extension of the Gini inequality index. *International Economic Review*, 24: 617–628.

Zellner, A. and Revankar, N.S. 1969. Generalized production functions. *Review of Economic Studies*, 36: 241–250.

Zheng, B. and Zhang, Y. 2013. Household expenditures for leisure tourism in the USA, 1996 and 2006. *International Journal of Tourism Research*, 15(2): 197–208.

Zipf, G.K. 1949. *Human behavior and the principle of least effort. Addison–Wesley Press.* Cambridge, MA.

Index

ACD model, 138, 140–147, 157

Characteristic function, 8
City size, 121, 130–136, 138, 139
Comparison of risks, 77
Composite models
    definition, 33
    general, 35
Compound model, 102, 104, 105, 107, 119
Convolution, 103, 104, 106
Correlation, 91, 102, 108, 110, 111, 119, 147
Cost frontier, 121–123, 125, 156
Covariate, 86, 90, 95–99, 101, 106, 107, 111–113, 116, 121, 122, 143
Cumulant generating function, 36, 52, 55
Cumulative distribution function, 16, 17, 19, 20, 22, 24–28, 33, 53, 55, 63, 67, 74, 76, 82, 103, 126, 133, 137

Distribution
    Bernoulli, 2–4, 7, 11, 51
    beta, 28
    bimodal, 5, 85, 90–94, 98
    binomial, 3–7, 9, 51, 55–57, 83
    Birnbaum-Saunders, 140
    bivariate, 86, 108, 113, 119
        exponential, 2, 41, 48
        normal, 38, 40, 48
        Poisson, 35–37, 48

Burr, 140, 143, 145, 147
composite, 132, 135, 138
    lognormal-Pareto, 34, 132, 133, 135, 137
conditional, 1, 21, 29, 30, 32, 39, 41, 47, 93, 111, 124–127, 156, 157
Conway-Maxwell-Poisson, 91
exponential, 20–23, 26, 32, 38, 41, 47, 52, 63, 69, 71, 74, 75, 78, 79, 83, 91, 94, 103, 104, 106, 118, 122–124, 129, 140–142, 144, 145, 147, 148, 157
exponential-inverse Gaussian, 32, 145, 147
Frechet, 23
gamma, 13, 18–20, 22–24, 30, 32, 93, 100, 104, 108, 118, 123, 140, 157, 158
generalized
    exponential, 142
    gamma, 22, 100, 140, 145, 147
    inverse Gaussian, 25
    Pareto, 27
    Poisson, 91
geometric, 11, 13–15, 46, 56, 57, 59, 94
half normal, 123, 124, 126, 129, 156
inverse Gaussian, 23–25, 30, 32, 100, 143, 144, 147, 157
Lindley, 157
log skew-normal, 100

log-logistic, 23
logarithmic, 15, 16, 57, 105
lognormal, 16–18, 23–25, 68, 130–133, 135–138, 140
Lomax, 27, 63, 84
marginal, 36, 37, 39, 41, 48, 108, 110, 124–127, 156, 157
mixture, 1, 13, 29, 33, 90–92, 104, 106, 132, 144, 147, 158
multimodal, 90, 91
multivariate, 1, 35, 38
   normal, 1, 38, 39
negative binomial, 6, 11–13, 30, 37, 49, 55–57, 60, 91, 93, 97, 106, 107
negative binomial-inverse Gaussian, 31
normal, 2, 16–18, 20, 24, 25, 34, 38–40, 68, 91, 94, 116, 122, 124–128, 156, 157
Pareto, 25–27, 34, 47, 84, 130–133, 135–138, 140
Poisson, 1, 6–11, 13, 30, 31, 36, 37, 47, 49, 55–58, 83, 90–95, 104, 107, 108
Poisson-inverse Gaussian, 31, 83
positive
   negative binomial, 102, 105, 107, 118
   Poisson, 102, 104, 105, 118
Rayleigh, 22
shifted
   geometric, 108
   Poisson, 94
skew-normal, 17, 126
truncated normal, 123–125, 127, 129
unimodal, 5, 18, 23, 24, 91, 132
Weibull, 21–23, 47, 140, 142, 145, 147
Duration, 23, 85, 91, 121, 138, 140, 143–147

Elasticity, 123

Esscher
   transform, 63
Expected Shortfall, 71
Expenditure, 85–87, 89, 90, 98–103, 107, 108, 111, 113, 116

Farlie-Gumbel-Morgenstern copula, 108, 113, 115
Function
   Bessel (first kind), 105
   Bessel (third kind), 25, 31, 47
   beta, 28
   confluent hypergeometric, 28, 106, 110, 157
   digamma, 112
   exponential integral, 158
   gamma, 12, 28, 109, 112
   hypergeometric, 111
   incomplete beta, 28
   incomplete gamma, 19, 109, 110

Generalized additive model, 116, 117
Geography, 121, 130

Hazard function, 22, 23, 144

Income, 87, 88, 90, 91, 98, 99, 107, 113, 115, 121, 147, 149–151, 153
Index
   Gini, 150, 151, 154, 158
   of dispersion, 95
   Pietra, 150, 151, 154, 158
   tail, 136
   Yitzhaki, 150, 151, 157
Inequality, 147, 150

Law
   Gibrat, 130, 136
   power, 130, 131, 135, 137, 138
   Zipf, 130, 132, 133, 135–139
Leimkuhler curve, 149, 155, 156
Length of stay, 85, 86, 88, 90–93, 96–99, 102, 105, 108, 111, 113, 115, 117

INDEX

Link
    log, 96, 107, 117
    logit, 96, 112
Lorenz
    curve, 122, 147, 149–151, 153, 154, 157
        Aggarwal, 155, 158
        Gupta, 153, 154
        Pareto, 151, 153, 154
        Power, 153, 154, 158
    ordering, 151, 158

Mode, 5, 90, 91, 113, 123
Moment generating function, 2, 4, 8, 12, 14, 16–19, 21, 22, 24, 26–28, 36, 37, 46, 51, 53, 63, 64

Ordinary least square, 91, 116, 117, 128–130
Overdispersion, 85, 91, 93, 95, 105, 147

Plots
    log-log plot, 44, 138
    Q-Q plot, 45, 46
Premium, 49, 50, 62, 63
    calculation principle, 49, 62, 65, 67
    Esscher, 63, 64
    expected value, 63, 66
    loading factor, 63
    standard deviation, 63, 66
    Stop-loss, 50, 70
    variance, 66
Premium calculation principle, 65
Probability
    density function, 1, 16–29, 32–35, 38, 40, 41, 47, 53, 61, 63, 64, 69, 78, 82, 84, 93–97, 100–103, 106, 109, 118, 119, 124–127, 132, 133, 140, 142, 143, 145, 151, 157, 158

    generating function, 4, 5, 8, 10, 12–14, 16, 36, 54, 94, 104, 105, 118
    mass function, 2, 3, 5–8, 11, 13, 15, 31, 35–37, 47, 56, 57, 60, 64, 93, 94
Production
    Cobb-Douglas, 121, 123, 127, 129
    frontier, 121–123, 127–130

Rank, 130, 137
Reinsurance, 73–75
    excess of loss, 75
    global type, 74
    individualized type, 74
    proportional, 74
    Stop-loss , 76
Risk measure, 49, 50, 67, 69, 71–73

Severity, 25, 69
Smooth term, 116
Statistic
    AIC, 44, 97, 100, 101, 107, 116, 141, 147, 148
    BIC, 44, 141, 147
    CAIC, 44, 100, 101, 107, 141, 148
    excess-dispersion, 146
    HQIC, 44, 136
    Ljung-Box, 147
    MAE, 147, 152, 154, 158
    MAX, 152, 154, 158
    MSE, 147, 154
    NLL, 57, 100, 101, 113, 116, 136, 148
    SSE, 152, 154, 158
Statistics
    Chi-square test, 42, 57
Stochastic
    frontier analysis, 121–124, 126, 129, 156, 157
    order, 67, 151
        convex order, 80
        stochastic dominance, 77
        stop-loss order, 79

Tail Value at Risk, 69, 70
    conditional, 70, 71
    exponential, 70, 72
Technical efficiency, 121–127, 130, 131
Test
    Anderson Darling, 100
    Cramér-Von Mises, 100
    Kolmogorov-Smirnov, 100, 136, 137
    Vuong, 43, 97

Uncomplete moment, 33

Value at Risk, 67, 68, 70, 73
    exponential, 70, 72